INTERNATIONAL
WILDLIFE
ENCYCLOPEDIA

THIRD EDITION

Volume 4

CHI–CRA

Marshall Cavendish Corporation
99 White Plains Road
Tarrytown, New York 10591–9001

Website: www.marshallcavendish.com

© 2002 Marshall Cavendish Corporation

Library of Congress Cataloging-in-Publication Data

Burton, Maurice, 1898-
    International wildlife encyclopedia / [Maurice Burton, Robert
    Burton] .-- 3rd ed.
        p. cm.
    Includes bibliographical references (p.).
    Contents: v. 1. Aardvark - barnacle goose -- v. 2. Barn owl - brow-antlered deer -- v. 3. Brown bear - cheetah -- v. 4. Chickaree - crabs -- v. 5. Crab spider - ducks and geese -- v. 6. Dugong - flounder -- v. 7. Flowerpecker - golden mole -- v. 8. Golden oriole - hartebeest -- v. 9. Harvesting ant - jackal -- v. 10. Jackdaw - lemur -- v. 11. Leopard - marten -- v. 12. Martial eagle - needlefish -- v. 13. Newt - paradise fish -- v. 14. Paradoxical frog - poorwill -- v. 15. Porbeagle - rice rat -- v. 16. Rifleman - sea slug -- v. 17. Sea snake - sole -- v. 18. Solenodon - swan -- v. 19. Sweetfish - tree snake -- v. 20. Tree squirrel - water spider -- v. 21. Water vole - zorille -- v. 22. Index volume.
    ISBN 0-7614-7266-5 (set) -- ISBN 0-7614-7267-3 (v. 1) -- ISBN 0-7614-7268-1 (v. 2) -- ISBN 0-7614-7269-X (v. 3) -- ISBN 0-7614-7270-3 (v. 4) -- ISBN 0-7614-7271-1 (v. 5) -- ISBN 0-7614-7272-X (v. 6) -- ISBN 0-7614-7273-8 (v. 7) -- ISBN 0-7614-7274-6 (v. 8) -- ISBN 0-7614-7275-4 (v. 9) -- ISBN 0-7614-7276-2 (v. 10) -- ISBN 0-7614-7277-0 (v. 11) -- ISBN 0-7614-7278-9 (v. 12) -- ISBN 0-7614-7279-7 (v. 13) -- ISBN 0-7614-7280-0 (v. 14) -- ISBN 0-7614-7281-9 (v. 15) -- ISBN 0-7614-7282-7 (v. 16) -- ISBN 0-7614-7283-5 (v. 17) -- ISBN 0-7614-7284-3 (v. 18) -- ISBN 0-7614-7285-1 (v. 19) -- ISBN 0-7614-7286-X (v. 20) -- ISBN 0-7614-7287-8 (v. 21) -- ISBN 0-7614-7288-6 (v. 22)
    1. Zoology -- Dictionaries.   I. Burton, Robert, 1941-   . II. Title.

    QL9 .B796 2002
    590'.3--dc21

                                                        2001017458

Printed in Malaysia
Bound in the United States of America

07 06 05 04 03 02 01   8 7 6 5 4 3 2 1

Brown Partworks

Project editor: Ben Hoare
Associate editors: Lesley Campbell-Wright, Rob Dimery, Robert Houston, Jane Lanigan, Sally McFall, Chris Marshall, Paul Thompson, Matthew D. S. Turner
Managing editor: Tim Cooke
Designer: Paul Griffin
Picture researchers: Brenda Clynch, Becky Cox
Illustrators: Ian Lycett, Catherine Ward
Indexer: Kay Ollerenshaw

Marshall Cavendish Corporation
Editorial director: Paul Bernabeo

Authors and Consultants

Dr. Roger Avery, BSc, PhD (University of Bristol)

Rob Cave, BA (University of Plymouth)

Fergus Collins, BA (University of Liverpool)

Dr. Julia J. Day, BSc (University of Bristol), PhD (University of London)

Tom Day, BA, MA (University of Cambridge), MSc (University of Southampton)

Bridget Giles, BA (University of London)

Leon Gray, BSc (University of London)

Tim Harris, BSc (University of Reading)

Richard Hoey, BSc, MPhil (University of Manchester), MSc (University of London)

Dr. Terry J. Holt, BSc, PhD (University of Liverpool)

Dr. Robert D. Houston, BA, MA (University of Oxford), PhD (University of Bristol)

Steve Hurley, BSc (University of London), MRes (University of York)

Tom Jackson, BSc (University of Bristol)

E. Vicky Jenkins, BSc (University of Edinburgh), MSc (University of Aberdeen)

Dr. Jamie McDonald, BSc (University of York), PhD (University of Birmingham)

Dr. Robbie A. McDonald, BSc (University of St. Andrews), PhD (University of Bristol)

Dr. James W. R. Martin, BSc (University of Leeds), PhD (University of Bristol)

Dr. Tabetha Newman, BSc, PhD (University of Bristol)

Dr. J. Pimenta, BSc (University of London), PhD (University of Bristol)

Dr. Kieren Pitts, BSc, MSc (University of Exeter), PhD (University of Bristol)

Dr. Stephen J. Rossiter, BSc (University of Sussex), PhD (University of Bristol)

Dr. Sugoto Roy, PhD (University of Bristol)

Dr. Adrian Seymour, BSc, PhD (University of Bristol)

Dr. Salma H. A. Shalla, BSc, MSc, PhD (Suez Canal University, Egypt)

Dr. S. Stefanni, PhD (University of Bristol)

Steve Swaby, BA (University of Exeter)

Matthew D. S. Turner, BA (University of Loughborough), FZSL (Fellow of the Zoological Society of London)

Alastair Ward, BSc (University of Glasgow), MRes (University of York)

Dr. Michael J. Weedon, BSc, MSc, PhD (University of Bristol)

Alwyne Wheeler, former Head of the Fish Section, Natural History Museum, London

Contents

CHICKAREE

The Douglas squirrel is one of the smallest North American tree squirrels. It is noisier and more active than its larger relatives.

THREE SPECIES OF TREE SQUIRRELS found in North America are known as chickarees because of the abrasive, drawn-out *churr-churr* calls that they utter when disturbed. They are no more than two-thirds the size of the stronger-bodied gray squirrel. The chickarees differ only slightly in coloring.

The red, or pine, squirrel, *Tamiasciurus hud-sonicus*, ranges from Alaska and Quebec south through the Rocky Mountains to New Mexico and through the Appalachians to South Carolina. Its coat is tawny to brownish, with a rufous band running from the ears to the tip of the tail, and a blackish line along the sides. Around each eye is a narrow ring of white. The underparts are also white. In winter the coloring is drabber, and the line along the side, which is absent in young animals, disappears. Black or reddish ear-tufts are often grown in winter. A common species of Eurasian tree squirrel, *Sciurus vulgaris*, is also called the red squirrel.

The Douglas squirrel, *T. douglasii*, is found from British Columbia south to California and is one of the smallest tree squirrels in its range. It resembles the North American red squirrel, but has grayish brown upperparts and a rusty white

tinge to its underparts. The third species of chickaree, Mearn's squirrel, *T. mearnsi*, is confined to Baja California. Some authorities regard this form as a subspecies of *T. douglasi*.

Adaptable squirrels

The main habitat of chickarees is evergreen forest but they also occur in deciduous and mixed woods. Chickarees are very adaptable, making their homes wherever there is shelter, including around houses, in hedgerows and underground. They dig tunnels or use those abandoned by groundhogs (woodchucks) and chipmunks. In the winter, tunnels may be made under snowdrifts. Because chickarees can live underground and under piles of stones and logs, they have generally survived the wholesale felling of forests that has reduced the ranges of other tree squirrels.

Chickaree nests may be made in natural cavities in old trees, in the forks of branches, among foliage, in deserted woodpecker or flicker holes, underground and under stones or stumps. Occasionally an old hawk or crow nest is used as the foundation for the squirrels' shelters. The nest is made of three layers. Outside there is a

CHICKAREES

CLASS **Mammalia**

ORDER **Rodentia**

FAMILY **Sciuridae**

GENUS AND SPECIES **Red squirrel,** *Tamiasciurus hudsonicus*; **Douglas squirrel,** *T. douglasii*; **Mearn's squirrel,** *T. mearnsi*

ALTERNATIVE NAMES
Chickaree; pine squirrel (*T. hudsonicus* only)

WEIGHT
5–11 oz. (140–310 g)

LENGTH
Head and body: 6½–15 in. (16.5–38 cm); tail: 3½–6 in. (9–15 cm)

DISTINCTIVE FEATURES
Long, bushy tail; large eyes; short, round, furred ears, tufted in some individuals

DIET
Conifer seeds, nuts, buds, fruits and fungi; also bird eggs and nestlings, small rodents and carrion

BREEDING
Age at first breeding: 10–12 months; breeding season: February–April and fall; number of young: usually 4 to 6; gestation period: 33–35 days; breeding interval: 2 litters per year

LIFE SPAN
Up to 7 years

HABITAT
Mainly coniferous forest; also deciduous and mixed woodland and parks

DISTRIBUTION
North America, from Alaska and Quebec south to Baja California and South Carolina

STATUS
Generally common; critically endangered: Mount Graham red squirrel, *T. h. grahamensis*

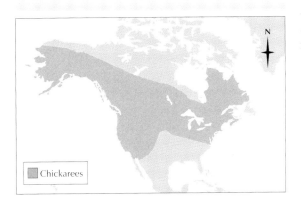

Chickarees

loose maze of leafy twigs. Inside is a compact, weatherproof layer of compressed leaves. The inner chamber is 4–6 inches (10–15 cm) across and lined with finer material, such as dried grasses, moss, feathers and fur.

Chickarees are diurnal (active during the day), but also emerge on moonlit nights. They are active throughout the winter, staying in their nests only during exceptionally bad weather.

Diet of pine cones

Chickarees appear to damage and reject more food than they actually eat, and store provisions that are never eaten. Their diet is mainly vegetarian, pine seeds being the staple throughout the year. Chickarees bite the green cones off the branches in the fall before they have time to ripen and disintegrate. The cones are carried

Chickarees feed mainly on pine seeds. Feeding activity peaks during the 2 hours immediately after sunrise and before sunset.

The red squirrel, like the two other chickarees, remains active all winter.

down the tree and stored in caches of 150 or more in damp soil so they do not ripen before the chickarees need them.

Many other seeds, nuts and berries are eaten in season. In spring chickarees tear up the buds of maple and elm trees and split the bark of maples to suck out the syrup. Apples are torn open and the seeds eaten, though the pulp is usually left. Chickarees also eat fungi, often storing it inside crevices in trees and stumps until it is dry. The bark of poplars and aspens are eaten in times of food shortage. Apart from vegetable matter, chickarees feed on insects, including beetle larvae dug out of timber, carrion and occasionally bird eggs and nestlings. Sometimes chickarees have been known to kill and eat young cottontail rabbits.

Teeth arrangement

All squirrels have a gap known as a diastema after their upper and lower incisors, in the place that canine teeth are found in carnivorous animals. Behind this gap, squirrels have cheek, or grinding, teeth, which consist of molars and premolars. As is the case with other rodents, the squirrels' incisors constantly grow to compensate for the continual wear that they receive because of the animals' mainly herbivorous diet.

Mating chases

Courtship starts in spring, or sometimes as early as January. As with other squirrels, the courting ritual involves a mating chase in which several chickarees chase each other about trees and over the ground, each one making a soft, coughing call. Eventually the chickarees pair off and mating takes place in February–April, depending on the climate. It is believed that both members of the pair help in building the nest, but the male takes no part in rearing the young.

The young, which usually number four to six, or rarely up to eight, are born blind and naked. They grow a fine covering of hair within 10 days. Young chickarees are nursed for 5–9 weeks, then at 18 weeks they disperse or are driven away by their mother, which may have a second litter in the fall.

Chickarees fall prey to bobcats, mink, the larger hawks and owls, but regionally their chief predators are several weasel-like mammals, the fishers and martens. These carnivores climb almost as well as the squirrels, and can follow them along the thinnest branches and leap from tree to tree. Large fur-bearing animals have become rare in many areas, and so chickarees are sometimes hunted by humans for their pelts, which are used as trimmings for coats and hats.

CHIMPANZEE

ONE OF THE GREAT APES, and the nearest in intelligence to humans, the chimpanzee has been studied in more detail than almost any other mammal. Yet despite extensive laboratory research into the chimp's capabilities, the species' wild behavior was not closely observed until relatively recently.

Being apes and not monkeys, chimpanzees lack a tail. Their arms are longer than their legs and their armspan is 50 percent longer than their height; the ratio is similar in humans. Their feet are narrow and have a large opposable "thumb," or big toe, which is used to improve grip when climbing and swinging through the treetops. Chimpanzees are robust, heavyweight animals and travel on the ground for up to 90 percent of the time. They normally move on all fours,

though they can also walk upright, with toes turned outward. When erect, chimpanzees stand 3⅓–5½ feet (1–1.8 m) tall. However, bipedal movement is strenuous for chimpanzees and they rarely sustain an erect posture for more than 150 feet (45 m). The hair of chimpanzees is long and coarse, and is black except for a white patch near the rump. The face, ears, hands and feet are all free from hair.

Two species of chimp
There are two species of chimpanzee: the more familiar common chimpanzee, *Pan troglodytes*, and the scarcer, much less widespread bonobo, *P. paniscus*. The bonobo is also known as the pygmy chimpanzee, although this name is misleading because the species is only slightly

Chimpanzees use hand gestures, facial expressions and about 35 different sounds to articulate their emotions, such as frustration (above).

CHIMPANZEE

smaller than the common chimpanzee. Both
chimpanzees live in the rain forests of West
Africa, especially in the lowlands of the Niger
Basin. The common chimpanzee also occurs in
montane forest, thin strips of forest (known as
gallery forest) and wooded savanna. Both species
forage in the forest canopy and each evening
make nests of branches and vines to sleep in.
However, they spend a considerable amount of
time on or near the ground.

Social mobility

Chimpanzees live in small groups, occasionally
of up to 40 animals. The usual size of a group is
three to six, although numbers increase when
chimps congregate to take advantage of a source
of plentiful and nutritious food, or when males
gather round a mature female on heat. A chimpanzee party constantly varies in size as various
members leave to wander in the forest by themselves or return from such an expedition. For this
reason, chimps are known as political animals.

*A young chimpanzee is
entirely dependent on
its mother for the first
2 years of its life.*

CHIMPANZEES

CLASS **Mammalia**

ORDER **Primates**

FAMILY **Pongidae**

GENUS AND SPECIES **Common chimpanzee,
Pan troglodytes; bonobo, P. paniscus**

ALTERNATIVE NAME
P. paniscus: pygmy chimpanzee

WEIGHT
**P. troglodytes: up to 90 lb. (40 kg).
P. paniscus: up to 75 lb. (35 kg).**

LENGTH
**P. troglodytes. Head and body: 26–37 in.
(65–95 cm); erect height: 3⅓–5½ ft. (1–1.8 m).
P. paniscus. Head and body: 28–33½ in.
(70–85 cm); erect height: 3⅓–5 ft. (1–1.5 m).**

DISTINCTIVE FEATURES
**Robust body; arms very powerful and longer
than legs; large opposable thumbs; black,
shaggy fur; bald face, ears, hands and feet**

DIET
**Mainly fruits and seeds; also leaves, roots,
invertebrates, bird eggs and nestlings and
mammals up to size of young bushbuck**

BREEDING
**Age at first breeding: 15–16 years (male),
13–14 years (female); number of young: 1,
rarely 2; gestation period: about 230 days**

LIFE SPAN
Up to 60 years

HABITAT
**Tropical rain forest; montane forest and
wooded savanna (P. troglodytes only)**

DISTRIBUTION
**P. troglodytes: southern Senegal east to
western Uganda and Tanzania. P. paniscus:
Congo and Democratic Republic of Congo.**

STATUS
Both species endangered

Chimpanzees

The only constant unit of social life among chimpanzees is a mother and her youngest infant. An adult female may have two or three young of different ages with her at any time because juveniles stay with their mothers for several years.

Within a chimpanzee party, the males are arranged in a social order. Dominance is related to age: a male chimpanzee gradually rises in

The bonobo

Bonobos have dark, sleek fur and proportionately longer limbs than common chimpanzees. They are also more arboreal (tree-living), spending much of their time in high tree branches. A bonobo community may spread over 8–25 square miles (20–60 sq km) of rain forest, large parties being attracted to prime feeding sites, such as

social position from the time he is physically mature and leaves the protection of his mother. The status of a male seems also to be partly determined by loud displays, such as charging while waving branches or rocks and drumming the feet on the proplike buttresses of forest trees. However, chimpanzees usually recognize the "right of ownership," to the extent that a dominant male will refrain from wresting food from one of his inferiors.

Alpha (top) males may remain dominant for up to 10 years, though they may be superseded by other individuals in a matter of weeks or months. Younger chimps ingratiate themselves with more dominant individuals by checking their fur and removing ticks and biting insects. Grooming defuses tensions within a group.

fruiting trees. Potential antagonisms between individual bonobos are defused by the species' promiscuous sexual behavior. Bonobos of both sexes and all ages participate in a wide range of sexual practices, suggesting that sexual activity has been diverted from a purely reproductive role in order to fulfill a social purpose. Scientists believe that seduction has overtaken male dominance as a regulatory factor in bonobo society.

Appetite for fruit

Chimpanzees spend about 7 hours a day feeding in trees or on the ground. They investigate any source likely to produce food. Crevices in logs are searched for insects and bird nests are robbed of eggs and chicks, but the chimpanzees' staple diet consists of fruits, leaves and roots. Crops of

Chimps use mutual grooming to defuse tensions and confirm their relationships with one another.

appear to know what to do with it, but by a combination of instinct and knowledge gained from having seen other mothers with their babies, she soon starts to nurse the infant. For 2 years the young chimp is completely dependent on its mother. At first the mother carries her baby to her breast, but as it grows larger it rides on her back. When a young chimp leaves its mother's back it is given considerable freedom within the group, and is able to climb over dominant males without retaliation.

Tool users

Chimps are the most advanced tool users in the animal world, apart from humans. In captivity they have been seen to balance boxes on top of one other in order to reach otherwise inaccessible bananas. Observations made by the primatologist Jane Goodall and other naturalists have shown that tool use is learned by imitation. Certain more "difficult" tools that require advanced dexterity and coordination may take several years to master, and are understood by only a few chimps in each area.

Chimpanzees are able to use a variety of tools, but some individuals master more tools than others. The most common task performed with the assistance of tools is the extraction of honey, ants and termites from nests. Sticks are picked off the ground or broken from branches, pushed into nests, withdrawn, and the honey or insects licked off. Stones are also used to crack nuts, and as missiles to drive humans and baboons away from food sources. The stones, which may weigh several pounds, are thrown overarm, not very accurately but definitely aimed. Some chimps have learned how to hold a leafy twig and use it as a fly whisk, brushing away biting insects. A particularly sophisticated instance of tool use is when chimps chew a mouthful of leaves to make an absorbent "sponge," which is repeatedly dipped into the water-filled cavity of a rotten tree trunk and removed to be sucked dry.

Wild populations in danger

In 1900 there were about 1 million chimpanzees, but by the late 1990s the wild population had fallen to less than 100,000. Chimpanzees are now extinct in several countries and endangered in all others. Young chimps are taken illegally for sale to private zoos and for use in scientific experiments. However, the most serious threats facing chimps are hunting for bush meat and the destruction of forests, especially in West Africa.

Many chimps know how to dip a stick into an ant or termite nest and then lick off the angry insects. Chimps are by far the most advanced tool users, after humans.

ripening fruit, such as bananas, pawpaws and wild figs, are a special attraction, and chimps can become a nuisance when they attack fruit plantations. A large male chimp is able to consume over 50 bananas at one sitting.

Hunting for meat

For many years it was thought that the only living creatures eaten by chimpanzees were insects and, occasionally, young birds and small rodents. However, zoologists have discovered that chimpanzees also hunt larger animals; it is now believed that meat constitutes about 2 percent of their diet. Chimpanzees can catch young bushbucks and bushpigs, as well as colobus monkeys and young baboons. They sometimes hunt in organized groups, with some group-members giving chase, some blocking the victims' escape routes, and others waiting in ambush. Cannibalism has also been observed in chimpanzees, although it is rare.

Care of young

Common chimpanzees are generally promiscuous, and when a female is receptive a crowd of excited males soon gathers round her. Most or all of the males, regardless of social standing, mate with the female. She remains on heat for several days, after which the males quickly lose interest. However, temporary pairs may form between a male and a female in oestrus.

The female chimpanzee gives birth to a single baby after about 230 days; twins are rare. If it is the female's first baby, she does not at first

CHINCHILLA

B EST KNOWN FOR THEIR remarkably soft fur, which is 1–1½ inches (2.5–4 cm) long, the chinchillas are rodents related to viscachas. They resemble small rabbits and may grow up to 15 inches (38 cm) in length, with a bushy, squirrel-like tail that may attain 6 inches (15 cm). A chinchilla's ears are almost hairless, the eyes are large and the whiskers are very long. The chinchilla uses its whiskers at night to gauge the width of rocky crevices that it seeks to enter. If the whiskers, which are as wide as the chinchilla's body, do not bend in the crevice, then the animal knows that there is enough space for it to enter. The small feet have weak claws on each of the four toes and the soles of the feet are rubbery, which provides the chinchilla with traction on the rocks of its montane habitat.

The fur is dense and silky, with 80 to 100 hairs growing out of each hair follicle. Originally chinchillas living in the wild had a mottled yellowish-gray coat. Today, most chinchillas have bluish-gray fur with faint dusky markings and their underside is whitish. Chinchillas must take regular dust baths in order to absorb the oils that naturally accumulate on their coats.

The cry of a chinchilla is a clicking sound, and is used from birth to signal to others. A female will often click at her young if she is nipped while feeding, or if they are fighting among themselves.

Life in high mountains

Most scientists recognize two chinchilla species. *Chinchilla laniger*, which has a long tail and ears, lives at relatively low levels in the Chilean Andes. The other species, *C. brevicaudata*, has a shorter tail and ears, and lives high up in the mountains in Bolivia, Argentina and northern Chile. A small population of *C. brevicaudata* may still survive in Peru, though many zoologists

Chinchillas live in groups in burrows or among rocks, but hunting has reduced their numbers to such an extent that large colonies are very rare.

believe that the chinchillas in this region are now extinct. The physical differences between the two chinchilla species may have arisen due to the differing climates of their respective ranges. Animals living in cold climates tend to have less pronounced physical extremities as such features increase their surface area, which leads to greater heat loss. As a result, chinchillas living at 15,000 feet (4,600 m) have smaller ears and shorter tails than those living at 5,000 feet (1,500 m).

Shared burrows

Chinchillas live communally in semiarid mountain areas, in burrows or in the crevices among rocks. They are nocturnal, though they will emerge to bask in the morning and evening sun. Before chinchilla numbers became depleted by excessive hunting and habitat destruction, some colonies contained as many as 100 animals.

The staple foods of chinchillas are coarse grasses, herbs, fruits, cacti, roots and the bark of shrubs. Food is held in the forepaws and eaten while sitting back on the haunches. There is little water in the cold, arid landscape that chinchillas inhabit and they must rely on water obtained from plants and dew.

Prolonged gestation

Chinchillas mate for life, and the female is the dominant member of the pair. She is larger than the male, and if there is any fighting, she generally wins. As part of the courtship ritual, the male and female chinchilla pull out tufts of each other's hair. Female chinchillas can be very aggressive toward one another and sometimes also show aggression toward male chinchillas, even when sexually receptive.

A chinchilla litter usually consists of two or three young, although litters of up to six occur. The gestation time is about 3½ months, which is unusually long for a small mammal; the mother weighs less than a European rabbit, but carries her young for four times as long. Such an extended gestation is advantageous in the chinchillas' bleak habitat. Young chinchillas enter the world having already passed the most helpless stages of infancy, arriving fully furred and thus not so liable to lose heat.

The young can run within hours of their birth but, for greater protection against the cold, nestle between their parents, which squat side by side to form a shelter. They begin to eat solid food after 1 week, but are not weaned until 7–8 weeks old. Juvenile chinchillas become sexually mature within a year, and one to three litters per year is usual in adult females. The typical life expectancy of a chinchilla is up to 10 years, although in captivity chinchillas are known to live twice as long.

CHINCHILLAS

CLASS	**Mammalia**
ORDER	**Rodentia**
FAMILY	**Chinchillidae**

GENUS AND SPECIES **Short-tailed chinchilla,** *Chinchilla brevicaudata*; **long-tailed chinchilla,** *C. laniger*

WEIGHT
1–1¾ lb. (500–800 g)

LENGTH
Head and body: 9–15 in. (23–38 cm); tail: 3–6 in. (7.5–15 cm); female larger than male

DISTINCTIVE FEATURES
Rabbitlike rodent; large eyes; prominent rounded ears; long whiskers; bushy tail; dense coat of silky, bluish gray fur

DIET
Montane vegetation including herbs, coarse grasses, cacti, roots, fruits and bark

BREEDING
Age at first breeding: 8 months; breeding season: all year; number of young: usually 2 or 3; gestation period: average 110 days; breeding interval: 1 to 3 litters per year

LIFE SPAN
Up to 10 years; up to 20 years in captivity

HABITAT
Semiarid, rocky areas in mountains; *C. brevicaudata* at higher altitudes

DISTRIBUTION
***C. brevicaudata*: Andes Mountains of western Argentina, Bolivia, northern Chile and Peru. *C. laniger*: Andes Mountains of Chile (formerly more widespread).**

STATUS
***C. brevicaudata*: critically endangered; near-extinct in Peru. *C. laniger*: vulnerable; total wild population less than 10,000.**

Chinchillas

Valuable pelts

Before Europeans reached South America, chinchilla pelts provided Native South Americans with material for warm clothes. The wealthier tribe members wore robes made of whole chinchilla skins, while the poor wove blankets from thread made of chinchilla hair.

Chinchilla fur did not become popular outside South America until the 18th century and it was only in the 19th century that demand began to threaten chinchilla populations. The peak of the commerce in chinchilla skins came in 1899, when half a million pelts were exported from Chile alone. In the early years of the 20th century chinchilla pelts became the most valuable pelts in the world in terms of size and weight: at one point, coats made of wild chinchilla fur were on sale for $100,000 each.

Legal protection

As a result of the international fur trade chinchillas became rare and in the 1900s South American governments imposed tariffs on the export of the animals' skins. However, this served to encourage smuggling, and it was soon necessary to ban the hunting and export of chinchillas. Pressure on the much depleted chinchilla populations was eased by the establishment of chinchilla ranches both inside and outside South America. Commercial farming caused the price of chinchilla pelts to fall, which in turn helped to protect wild stock by making hunting uneconomical. Today chinchillas enjoy full legal protection, although enforcement of the laws is difficult in remote areas. Chinchillas have been legally protected in Chile since 1929, but despite this have been hunted to near extinction and now exist there only in the coastal mountains. The ranges of both chinchilla species have contracted hugely and are now fragmentary.

Conservation measures

Chinchillas have become extremely popular as pets, and are more numerous in captivity than in the wild. Half of all wild chinchillas live within a fenced federal reserve in Auco, Chile. The second largest wild population is on private land in Coquimbo, Chile. Attempts have been made to reintroduce captive-bred chinchillas to the Andes Mountains, but so far without success.

Apart from hunting, chinchillas are threatened by changes to their natural habitat. The algarrobilla shrub on which they feed is burned and harvested and the fragile mountain fauna is damaged by overgrazing by domestic livestock. Other threats include wood collection, mining and competition for food and burrow sites from several other South American rodents, including the degu, *Octodon degus*, and chinchilla rats, in the genus *Abrocoma*.

Wild chinchillas have bluish gray fur, but captive animals have been bred to produce fur in various shades of silver, gray, black, white and brown.

CHINESE WATER DEER

Male Chinese water deer lack antlers and instead have a pair of unique, tusklike upper canine teeth.

A VERY SMALL DEER native to China and Korea which is better known in captivity than in the wild, the Chinese water deer is about 3 feet (91 cm) long, stands 19–21 inches (48–53 cm) at the shoulder and weighs 33–39½ pounds (15–18 kg). The coat is a light yellowish brown to pale reddish brown in summer, becoming dark brown in winter; the deer's tail is a hairy stump. The underparts and a narrow perpendicular band on the muzzle are white.

Deer with tusks

There is little difference in appearance between the sexes of the Chinese water deer. The males do not have antlers, but their upper canines are long and tusklike, their points reaching well below the lower jaw. Another peculiarity is a small scent gland in the groin area. Musk deer also have a musk gland, sometimes called an inguinal gland, in the same place.

The Chinese water deer is unique among deer in that the doe may give birth to up to six fawns, although two to three is more usual. However, in common with most deer she has only four teats. The breeding season is in autumn and early winter, when there is much fighting between bucks. They deliver slashing cuts with their tusklike canines, inflicting serious wounds, although these are seldom fatal. The young are born in June and July.

Rabbitlike deer

The largest populations of Chinese water deer are found in marshy beds in the river valleys of the Yangtze River basin. The species usually frequents long grass and tall reed beds fringing riverbanks, though it has also been sighted on mountainsides and farmland. In captivity the deer thrives without access to water, although it will enter pools or ponds if they are available. A darker colored subspecies lives in Korea.

The Chinese water deer seems to require long vegetation, into which it scuttles or drops when alarmed. The deer lives in small groups, which have an unusually closely knit social structure. In captivity, when one deer is removed from the group for a while the others refuse to accept it on its return.

CHINESE WATER DEER

CLASS **Mammalia**

ORDER **Artiodactyla**

FAMILY **Cervidae**

GENUS AND SPECIES *Hydropotes inermis*

ALTERNATIVE NAMES
Korean water deer; water deer; Chinese river deer; Yangtze River deer; river deer

WEIGHT
**Male: 37½–39½ lb. (17–18 kg).
Female: 33–35 lb. (15–16 kg).**

LENGTH
Head and body: about 36 in. (90 cm); shoulder height: 19–21 in. (48–53 cm); male larger than female

DISTINCTIVE FEATURES
Very small; lacks antlers but male instead has large, extended tusks (upper canines); proportionately large, rounded ears

DIET
Mainly grasses and herbs

BREEDING
Age at first breeding: 5–6 months (male), 7–8 months (female); breeding season: December; number of young: usually 2 or 3; gestation period: 170–180 days; breeding interval: 1 year

LIFE SPAN
Up to 10 years

HABITAT
Reed swamps and grasslands along lower stretches of large rivers

DISTRIBUTION
Eastern China and Korea; introduced to parts of Britain and France

STATUS
Not known, but population of more than 10,000 in Lower Yangtze River alone

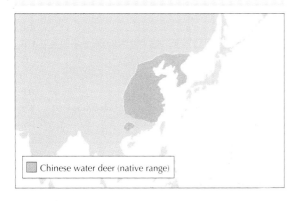

Chinese water deer (native range)

Chinese water deer appear to feed mainly on grasses and herbs. However, studies made of water deer in captivity suggest that this diet is probably supplemented by other plants.

Feral populations in Europe

Western scientists first became aware of Chinese water deer in 1870, and in 1873 a live specimen was received at London Zoo. Others followed in succeeding years and, in 1929–1931, 32 deer were taken to England, to Whipsnade and Woburn Park, which belonged to the Duke of Bedford. They were kept in semicaptivity and thrived. During the early 1940s some deer escaped from both places and began to spread into neighboring counties. In 1944 a few of the deer were sent from Woburn to private parks in Hampshire, in the south of England, and in 1950 and 1954 more were sent north to Yorkshire. Deer escaped from all of these sites, and as a result established feral populations in parts of southern England. Chinese water deer are adept at keeping out of sight, and consequently their numbers are difficult to assess.

Local superstitions arise from the ease with which Chinese water deer can escape observation. It is said that they were left unmolested in China because of local tradition, and in Korea their bite is believed to be fatal. People who keep water deer in parks in England testify to the need for avoiding the teeth of males.

Female water deer usually have two or three fawns, though litters of up to six have been recorded; more than in any other species of deer.

CHIPMUNK

The eastern chipmunk (above) is larger than the various species of western chipmunks, though its tail is proportionately shorter.

THERE ARE TWO GENERA of chipmunks, *Tamias* and *Eutamias*, among the many kinds of ground squirrels. *Tamias* contains a single species, the eastern chipmunk, *T. striatus*, which is the largest of all the chipmunks. Its fur is reddish brown, with dark stripes on the back, alternating with two lighter stripes. The tail is not as bushy as that of tree squirrels. *Eutamias* contains 17 species, collectively known as western chipmunks. All are smaller than the eastern chipmunk, but they have a similarly sized tail, which is therefore larger in relation to the body. Their fur is lighter and there are five light stripes between the dark stripes on the back. The eastern chipmunk has one upper premolar (grinding tooth) on each side of the jaw, whereas western chipmunks have two. The eastern chipmunk's rufous rump also distinguishes the two genera.

Variety of species

The designations eastern and western refer to the species' distribution in North America. The eastern chipmunk is widespread in the eastern United States north of Florida and Louisiana and also occurs in southeastern Canada. It thrives in forests and shrubland, and frequents fallen logs, rocks and outbuildings.

All but one of the western chipmunks are native to western and central North America, ranging from the Yukon to Sonora in Mexico. The remaining *Eutamias* species, the Siberian chipmunk, *E. sibiricus*, ranges from Siberia to northern China, Korea and northern Japan. Western chipmunks are found in a variety of forest and woodland, as well as in juniper scrub, chaparral and rocky areas. Some species are widespread, while others have very specialized habitat requirements and occur in small ranges.

Burrow systems

Chipmunks are skillful climbers, but prefer to stay on the ground, although the eastern chipmunk sometimes rears its young in trees. They make a complicated system of burrows underground, often running under logs and stones, or delving several feet under the turf. Each burrow

CHIPMUNKS

CLASS	**Mammalia**
ORDER	**Rodentia**
FAMILY	**Sciuridae**

GENUS AND SPECIES **Tamias: eastern chipmunk, T. striatus. Eutamias: 17 species, including Townsend's chipmunk, E. townsendii; least chipmunk, E. minimus; and Siberian chipmunk, E. sibiricus**

LENGTH
Tamias. Head and body: 5½–7½ in. (14–19 cm); tail: 3¼–4¼ in. (8–11 cm). Eutamias. Head and body: 3¼–6¼ in. (8–16 cm); tail: 2½–5½ in. (6–14 cm).

DISTINCTIVE FEATURES
Pointed head; prominent ears; bushy tail; brown-gray fur; black and white stripes on back, extending over eyes in Eutamias

DIET
Mainly seeds, fruits, fungi and grasses; also bird eggs, invertebrates and small vertebrates

BREEDING
Age at first breeding: 1 year; breeding season: February–April and June–August (Tamias), May–August (Eutamias); number of young: usually 4 or 5; gestation period: about 31 days; breeding interval: 1 or 2 litters per year (Tamias), 1 litter per year (Eutamias)

LIFE SPAN
Usually 2–3 years, sometimes up to 8 years

HABITAT
Forest, especially coniferous; also open woodland, scrub, chaparral and rocky scree

DISTRIBUTION
Tamias: eastern U.S. and southeastern Canada. Eutamias: western Canada south through western U.S. to Mexico; Siberia, Mongolia, Korea and Japan (E. sibiricus).

STATUS
Common, sometimes very common

Chipmunks

is owned by one chipmunk which continues digging throughout its life, so that the burrows may reach lengths of 30 feet (9 m) or more. The burrows may have more than one entrance and perhaps several side chambers, one of which may contain a nest of leaves and grasses.

Chipmunks do not hibernate in the strict sense of the word. However, during prolonged spells of bad weather the animals do enter a state of torpor, from which they awake from time to time to feed from caches (stores) of food.

Cheek pouches

Chipmunks feed mainly on berries, fruits, nuts and small seeds, which are collected after they have fallen to the ground, or which the chipmunks climb trees and shrubs to pick. Fungi, grasses and leaves are also eaten. Chipmunks are sometimes carnivorous, eating slugs, snails and insects. Bird eggs, mice and small snakes are also taken, and in the Sierra Nevada the eastern chipmunk is considered to be one of the chief predators of the rosy finch, *Leucosticte arctoa*.

Food that is not immediately needed is carried in the cheek pouches and cached for use in the winter. The cheek pouches are loose folds of skin, naked inside but not moist, that open into the side of the mouth. To fill its pouches, a chipmunk holds a nut in its paws, neatly bites off

Although adept at climbing, chipmunks spend most of their time near the ground. These are young Townsend's chipmunks.

During the fall the least chipmunk, like all chipmunks, devotes a lot of energy to gathering the hoards of food that will last it through the winter.

the sharp point on each end and slips it into one pouch. The next nut is placed on the other side and the pouches are filled alternately. The chipmunk can take up to four nuts in each pouch and another between the teeth.

Expert hoarders

Chipmunks are probably the most accomplished hoarders in the squirrel family, but are selective about the items they hoard. When a chipmunk is collecting its winter store, it selects only nuts, seeds and cones, never any fruit or flesh that would decompose.

The sheer bulk of chipmunk stores is highly impressive. There are reports of caches containing 8 quarts (9 l) of acorns or 32 quarts (35 l) of nuts, and one cache does not constitute a chipmunk's complete winter store. More than one cache may be made in the burrow, and small caches are made in secure hiding places across the chipmunk's home range. In this respect the chipmunk combines the hoarding behavior of the chickaree, which creates one or two large stores, with that of the gray squirrel, which usually makes many small ones.

Like the gray squirrel, the chipmunk sometimes forgets the location of some of its small, scattered stores. These often consist of just a mouthful of nuts buried under leaves or turf. During the winter the chipmunk may find some of the stores by smell; otherwise they remain hidden until they germinate and contribute to the growth of the woodland.

Breeding and predators

Mating takes place during the spring and summer, the males seeking out the females in their burrows. The young, usually four or five in number, are born about 1 month later. They spend a month in the nest, then begin to accompany their mother on foraging trips above ground, venturing further afield each time. The young remain as a group for 6 weeks before starting to forage independently. Eastern chipmunks may produce two litters in a season; western chipmunks only ever produce one.

Chipmunks are a major prey item of coyotes, bobcats, foxes, weasels, hawks, owls and snakes. When danger threatens, chipmunks give a scolding or whistling alarm call, and dash to cover.

CHOUGH

THE NAME CHOUGH COVERS three species belonging to the crow family. The red-billed or common chough, *Pyrrhocorax pyrrhocorax*, has a glossy blue-black plumage, usually with a greenish tinge on the wings and tail, as in the black-billed magpie, *Pica pica*. The adult is readily recognizable by its long, decurved red bill and red legs. Its call resembles the *caw* of another Eurasian crow, the jackdaw, *Corvus monedula*, but is higher and more musical. The old pronunciation of the chough's name, "chow," was an imitation of this call. The alpine or yellow-billed chough, *P. graculus*, closely resembles the red-billed chough, but has a shorter, yellowish bill and a different call, a shrill, rippling cry.

The Australian white-winged chough, *Corcorax melanorhamphos*, belongs to the magpie lark family, Grallidae. It occurs in scrub and dry forest in eastern Australia. The white-winged chough has a curved bill like the true choughs of Eurasia, and shows white on its primary wing feathers when the wings are spread.

Mountain crows

Red-billed and alpine choughs live in mountainous regions, but while the latter is strictly a montane species, the former also occurs at lower altitudes. In most countries where both species are found the alpine chough is more numerous. In Europe, the red-billed chough is found along the western coasts of Britain and Ireland, in Brittany in northwestern France, in much of Spain and in scattered parts of the Alps, mainland Italy, Sardinia, Sicily and the Balkans. A single isolated colony in Ethiopia, 1,500 miles (2,750 km) from any other, is probably a relict from an ice age during which the entire population of the species moved south. The alpine chough is found in many parts of the Alps, northern Italy, Corsica, the Balkans and the Spanish Pyrenees. Both species of true choughs also occur in Central Asia and across southern Asia to eastern China.

Changing fortunes

The red-billed chough was formerly known as the Cornish chough, a reference to the fact that it was once common on the sea cliffs of Cornwall in southwestern England. The species was last recorded there in 1952, and has also suffered long-term declines in the rest of Britain and Europe, although these have been reversed to some extent in recent years. The drop in European red-billed chough populations was caused largely by changes in agricultural practices, which reduced the availability of invertebrate prey. Also to blame was strong competition for food and nesting sites from the much more common and adaptable jackdaw.

On top of the world

Choughs seem to thrive at high altitudes. They regularly breed at 17,000 feet (5,200 m) and pairs have nested at up to 19,000 feet (5,800 m) in the

Unlike nearly all other crows, the choughs have brightly colored bills and feet: red in the red-billed chough (above) and yellow in the alpine chough.

Red-billed choughs feed primarily on insects but, like many crows, are great opportunists, seizing lizards and other large prey when available.

Himalayas. Mountaineers have seen choughs flying above the highest slopes of Mount Everest in Tibet. Survival in such rarefied (oxygen-poor) air must present choughs with problems, since birds require large amounts of oxygen for flight.

Sociable scavengers

Choughs feed largely on insects, which they reach by thrusting their curved bills into the soil. Important prey items include caterpillars, wireworms, ground beetles and ants, which are occasionally supplemented with larger prey. Some plant material is eaten in winter. Flocks of choughs often quarter the ground in long, bounding hops, and associate with other species of crows, including the jackdaw, rook (*Corvus frugilegus*) and carrion crow (*C. corone*). The alpine chough, in particular, scavenges rubbish dumps and ski resorts for anything edible, and often becomes quite tame.

Chough nests are loose constructions of twigs, heather, bracken and other materials, often including sheep wool. The typical site is an inaccessible ledge high up on a cliff wall or in a quarry. In Asia the red-billed chough favors buildings and nests in the Great Wall of China. In many places choughs breed in colonies.

The clutch consists of three to five eggs, sometimes six, in both species. Incubation begins when the first egg is laid, so the chicks hatch out at intervals. The female incubates the eggs alone, and is fed during this time by her mate. The young stay in the nest for up to 40 days, and are fed by the parents, which forage together. The young birds spend a few days near the nest, then begin to fly and are taken on foraging trips by the parents, during which they learn to hunt.

RED-BILLED CHOUGH

CLASS	**Aves**
ORDER	**Passeriformes**
FAMILY	**Corvidae**
GENUS AND SPECIES	***Pyrrhocorax pyrrhocorax***

ALTERNATIVE NAMES
Common chough; Cornish chough (archaic)

WEIGHT
9–12½ oz. (260–350 g)

LENGTH
Head to tail: 15½–15¾ in. (39–40 cm); wingspan: 29–35 in. (73–90 cm)

DISTINCTIVE FEATURES
Adult: red, slightly decurved bill; red legs; glossy blue-black plumage with green tinge to wings and tail. Juvenile: yellowish red bill.

DIET
Soil-dwelling invertebrates and a few small vertebrates; grain and berries in winter

BREEDING
Age at first breeding: 2–4 years; breeding season: April–August; number of eggs: usually 3 to 5; incubation period: 17–18 days; fledging period: 30–40 days; breeding interval: 1 year

LIFE SPAN
Up to 20 years

HABITAT
Coastal cliffs and inland crags in western Europe; mountains farther east

DISTRIBUTION
Rocky coasts in western British Isles and northwestern France; montane regions in central and southern Europe, northwestern Africa and Central Asia and from northern India north to southeastern Russia

STATUS
Locally common. Population: 20,000 to 75,000 pairs in Europe.

Red-billed chough

CHUCKWALLA

In common with many other lizards, chuckwallas shed their tail as a defense mechanism when threatened. The replacement tail has different markings than the original.

CHUCKWALLAS ARE PLUMP LIZARDS, usually 8–10 inches (20–25 cm) long, though they may grow up to 18 inches (45 cm). They weigh about 3⅓ pounds (1.5 kg) when mature. The color is variable, though most chuckwallas are predominantly brown or black. Just after the annual molt, the skin is shiny. Along the back run lines of dark brown, which continue down the tail. As the males grow older, these brown lines disappear and the body becomes lighter; the tail becomes almost white. These differences help to distinguish male and female chuckwallas, but it is still easy to confuse the sexes because young males look like females and the largest females resemble males.

Desert residents

There are several species of chuckwalla, and all of them live in desert regions. One species occurs in the southwestern United States, in areas of rock and lava in Utah, Arizona and New Mexico, and southward into northernmost Mexico. The other species are found only in Mexico. Chuckwallas keep within a home range but do not have regular "homes" or retreats. When resting or in times of danger, chuckwallas retire to any available rock crevice. The ranges of male chuckwallas incorporate several female ranges and never overlap, each male defending his range against other males. One study showed that male ranges are three times larger than those of females.

Summer sleep

Chuckwallas are active for only part of the year, from late March to early August, when food is most plentiful. Later in the year temperatures rise, the humidity drops due to the lack of rain and plants become dormant or wither. Faced with shortages of food and water the chuckwallas disappear below ground and estivate (pass the summer in a torpid state). They do not emerge even when conditions become temporarily favorable. The estivation, or summer sleep, becomes hibernation, or winter sleep, as the vegetation remains parched all winter.

Chuckwallas are active during the day, in air temperatures of 70–105° F (20–40° C). They are most active early in the day, before the air becomes too hot, and spend the first daylight hours basking. Chuckwallas can control their body temperature to some extent while lying in the sun. When the sun becomes too hot, the lizards move to the shade or lie in such a way that less of the body is exposed to the sun. The body color also becomes lighter, so that some of the sun's rays are reflected instead of being absorbed.

Flower-eaters

The staple food for most lizards is small animals, especially insects, but chuckwallas have been seen to feed on flesh only in captivity, where they eat mealworms and mice. In the wild chuckwallas eat plants, favoring blossoms, such as those of the prickly pear and creosote bush, rather than leaves, fruits and shoots.

When a chuckwalla emerges in the morning it basks for a while until its body has warmed up and it can become active. It then forages around its home range, wandering from plant to plant nibbling at the flowers, and sometimes also the leaves. Chuckwallas seem to prefer yellow petals. They browse low-growing plants and climb up higher shrubs, clambering around the outside to reach the flowers.

One species of chuckwalla, S. varius, is restricted to San Esteban Island in the Sea of Cortez, Mexico. It is now endangered.

CHUCKWALLAS

CLASS	**Reptilia**
ORDER	**Squamata**
SUBORDER	**Sauria**
FAMILY	**Iguanidae**

GENUS AND SPECIES **Several species, including *Sauromalus obesus*; and San Esteban Island chuckwalla, *S. varius***

ALTERNATIVE NAME
Chuck

LENGTH
Up to 18 in. (45 cm), usually 8–10 in. (20–25 cm)

DISTINCTIVE FEATURES
Plump, rather flat body; loose folds of skin on neck and sides of body

DIET
Mainly flower blossom; some leaves

BREEDING
Breeding season: spring and early summer; number of eggs: up to 10; breeding interval: 1 year (male), usually 2 years (female)

LIFE SPAN
Not known

HABITAT
Rocky places in deserts and semiarid country

DISTRIBUTION
S. obesus: southwestern U.S. and northern Mexico; other species in Mexico

STATUS
Generally common; endangered: S. varius

Multiple mating

Breeding takes place in the spring and early summer so that the young chuckwallas will have a plentiful supply of food. Courtship has not been observed, but the peak of mating probably occurs in May. Each male mates with the females within his territory. Females will mate only with the resident male so long as he is able to repel male intruders. Males are fertile every year, but it appears that females produce eggs in only alternate years. This may be because flowers are not very nutritious, and because in such an arid habitat it is difficult for a large lizard to build up body reserves sufficient for forming eggs.

A female chuckwalla in captivity was seen to dig a hole 4 inches (10 cm) wide horizontally into a bank of sand, to a distance of 15 inches (38 cm). As she made progress, the entrance became blocked with the sand that had been pushed back. After laying the eggs, the female turned and scraped this sand back over the clutch.

Saved by a tight squeeze

Chuckwallas are vulnerable to predators when they are feeding, especially if surprised when climbing up bushes. Their main predators are probably hawks.

On being surprised, a chuckwalla rushes to the nearest rock crevice, where it wedges itself in by its toes. The lizard is difficult to dislodge because when pulled backward, its body scales catch on the rock. The tail is folded back out of the way, but if the lizard is touched it lashes out; this may deter the predator. Alternatively, the predator may take the tail, leaving the rest of the lizard behind. In common with many other lizards, the chuckwalla is able to regenerate its tail. To increase its hold in the crevice the chuckwalla inflates its body. An inflated chuckwalla

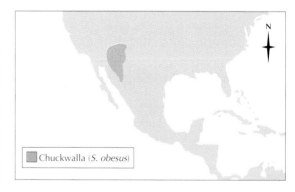

Chuckwalla (*S. obesus*)

increases its body volume by over 50 percent and so can exert considerable force against the walls of its sanctuary. Native Americans used to consider chuckwallas a delicacy, and to extract the lizards would stab them with a sharp stick, thus deflating the lungs.

CICADA

SEVERAL CICADAS ARE AMONG the largest of all insects. The adult Malaysian empress cicada, *Pomponia imperatoria*, has a wingspan of 8 inches (20 cm). Apart from their size and impressive appearance, cicadas are remarkable for their loud songs and for the extended period that some species take to mature.

The family Cicadidae, classified in the suborder Homoptera of the order Hemiptera, is related to the aphids, frog-hoppers and scale insects. Cicadas have a broad, flattened body and two pairs of large wings with a characteristic pattern of veins. The forewings overlap the hind wings and help to protect them. The longitudinal veins stop short of the edge of the wing, leaving a narrow, uninterrupted margin along the outer border. Generally the wings are transparent but in some species they are colored and patterned, the wing membrane being pigmented. It follows that the color cannot be brushed off as it can in butterflies and moths, the wings of which are covered with loosely attached scales.

By far the greatest diversity of cicadas is found in tropical regions, though the insects occur in temperate climates in small numbers. A few species occur in southern Europe and about 75 species are present in North America. More than 1,500 species are known altogether.

Loudest insects

Adult cicadas spend much of their time sitting high up on the trunks of trees or among the branches and foliage, singing intermittently. In almost all cicadas it is only the males that sing, though in a few species both sexes are vocal. Male cicadas produce a louder song than any other insect and can be heard up to ¼ mile (400 m) away. Some species sing during the day, while others sing only at dusk or dawn.

The purpose of the cicadas' songs is to attract mates. Both male and female cicadas have hearing organs and can recognize the distinctive song of their own species. As with the songs of frogs and birds, it highly likely that the strongest

Male cicadas produce a monotonous shrill whistle to attract females. Their chorus is sometimes loud enough to drown out everything else.

singers are able to attract the greatest number of mates, which also tend to be the fittest and strongest partners available.

Vibrating membranes

The singing apparatus of cicadas consists of a pair of membranes at the base of the abdomen, each surrounded and held by a stiff elastic ring. Each membrane is convex when relaxed, but an attached muscle can pull it down and then allow it to pop back, rather as a distorted tin lid can be popped in and out. In cicadas the membranes, also known as tymbals, oscillate at a rate from over 100 to nearly 500 times a second.

Several other muscles attached to the elastic ring are able to distort its shape, affecting both the volume and quality of the sound produced.

The entire singing apparatus is enclosed in a pair of air sacs, which act as resonating chambers to amplify the sound. These air sacs open and close to varying degrees to further modify the sound.

By means of this sophisticated vocalizing apparatus, each species of cicada produces its own unique signature tune. With experience humans can identify certain cicada species by their song alone.

When the adult cicada emerges from its larval case, it is pale and cannot fly. It soon turns a darker color and is able to use its wings.

CICADAS

PHYLUM	**Arthropoda**
CLASS	**Insecta**
ORDER	**Hemiptera**
SUBORDER	**Homoptera**
FAMILY	**Cicadidae**
GENUS	**Many, including *Magicicada*, *Tibicen*, *Cicadetta*, *Lyristes* and *Pomponia***
SPECIES	**More than 1,500, including periodic cicada, *M. septemdecim*; dogday harvestfly, *T. canicularis*; *C. montana*; *L. plebejus*; and Malaysian empress cicada, *P. imperatoria***

ALTERNATIVE NAME
M. septemdecim: 17-year cicada

LENGTH
½–4 in. (1–10 cm)

DISTINCTIVE FEATURES
Adult: broad, blunt head; wings large, held over body and usually transparent; most species brown, gray, green or blackish but some tropical species brightly colored. Larva: massive front legs; wings absent.

DIET
Adult: succulent shoots and sap of plant stems. Larva: sap of plant roots.

BREEDING
Varies according to species. Larval period: usually up to several years; 17 years (*M. septemdecim*).

HABITAT
Adult: trees and shrubs in wide variety of habitats. Larva: lives underground in soil.

DISTRIBUTION
Virtually worldwide; most species in Tropics, several species in warm temperate regions

STATUS
Most species common or abundant; 3 species of *Magicicada* threatened

Sweet rain

Like all members of the order Hemiptera, cicadas have mouthparts adapted for piercing and sucking, and they feed on the sap of plant stems and succulent shoots. Most of this sap is sugar and water, so cicadas must suck up large quantities to make an adequate meal. As a result, large amounts of a weak sugary solution are rapidly excreted. If it appears to be raining under a tree in a tropical forest when the sky above is clear, cicadas are probably feeding overhead. The "rain" consists of drops of sugary water excreted by the mass of hidden insects.

Underground larvae

The female cicada lays her eggs in slits in the twigs of trees and after hatching several weeks later the wingless larvae, or nymphs, fall to the ground. They dig down into the soil with their powerful front legs and quickly disappear underground. For the next period of their life the nymphs live in the soil, extracting sap from roots using the same method as adult cicadas. As the sap is very watery and does not contain much food, it is often some time before the nymphs attain full size and mature into adults. After a variable period of feeding, each nymph digs its way to the surface and climbs a tree, where it

rests. The skin of the larval case eventually splits, and the winged adult insect emerges. In some species of cicada the nymph builds an earthen cone projecting several inches above the ground, in which it remains for a time before its final transformation. The newly emerged adult is vulnerable to predators at first, as it is unable to fly, though it quickly gains the use of its wings.

Larvae that live for 17 years

In the North American periodic cicada, *Magicicada septemdecim*, the nymph spends up to 17 years underground in the course of its development, with 13 years being more typical in the southern parts of the species' range. The insect is also known as the 17-year cicada, though this does not mean that it appears in a district only once in 17 years, as there are often several broods in different stages of development at any one time. Major emergences of adult periodic cicadas, one of which took place in 1987, attract large numbers of predators, especially birds.

During its unusually long larval period, the periodic cicada passes through 7 instars, or molting stages. The extended development serves as a defense strategy because it helps prevent insect predators and parasites synchronizing their lifecycle with that of the cicada.

In tropical forests swarms of cicadas feeding in the treetops excrete a constant stream of sugary water droplets.

CICHLID

A red devil cichlid, Cichlasoma citrinellum, guarding its young. In many species of cichlid, both sexes look after the fry, taking them into the mouth if they are in danger.

THE 1,500 SPECIES OF CICHLID fish live in rivers and lakes in Africa, Madagascar and most of South and Central America north to Texas. There are also two species in southern India and Sri Lanka. Their bodies are flattened from side to side, as in the freshwater angelfish, *Pterophyllum scalare*. Most cichlids are strikingly colored, with a well-developed head and strong jaws, the lower of which juts out.

Strongly territorial

With the onset of breeding conditions, cichlids' colors become heightened, especially in the male. At the same time there is a noticeable difference in behavior. The male becomes far less sociable and lays claim to a part of the lake or river as his territory. Should another fish, especially one of his own species, swim into this region, the breeding male immediately goes into display. His colors become still more intense, his fins are fully erected, and the gill-covers are raised.

If the intruder is another male, subsequent action will depend upon whether he is in breeding condition or not. If he is not, he turns and flees, and is chased to the boundary of the territory. But an aggressive intruder will return the defending male's display. The two fish circle head to tail, each presenting a flank to the other, at the same time beating towards the other's flank with the tail. Very often there are no further developments, and the fight is broken off by the intruder's flight. However, the confrontation may result in the two contestants seizing each other by the mouth. In such cases, the owner of the territory is almost invariably victorious.

In many animal contests there seems to be a respect for ownership, with the owner of a territory winning the contest. The resident is aware of the value of the territory in terms of its capacity to provide food and shelter and of the effort invested to obtain it in the first place. The owner therefore has much more to lose than the intruder, which cannot know the precise value of the territory and will be less inclined to commit itself to fighting for an unknown reward.

Violent courtship

When a female cichlid of the same species swims into a male's territory, he gives the same reaction as for an intruding male: he displays belligerently. However, the female reacts by adopting an attitude of symbolic inferiority. Instead of retaliating she accepts the blows, which gradually subside as the male becomes aware of the presence of a potential mate. In some cichlids there are further preliminaries, the pair seizing each other by the mouth, tugging and twisting in an apparent trial of strength that may be repeated several times. Usually the struggle results in a successful mating, although it may end fatally for one or the other.

The choice of partner is only the first stage in courtship, during which the two fish do their best to guard the boundaries of their territory. The male does most of the displaying or fighting, the female assisting when necessary. Defending the territory is merely a means to an end, and preparations for spawning continue. These include digging pits in the sand with the mouth, and cleaning an area for reception of the spawn. The spawning site may also be the surface of a stone or, when cichlids are in an aquarium, part of the glass surface. Whatever site is chosen, the two fish set about cleaning it scrupulously with their mouths.

Parental protection

When the eggs are laid and fertilized, both parents take part in their care. One parent takes up a position over the surface on which the eggs

CICHLIDS

CLASS	Osteichthyes
ORDER	Perciformes
FAMILY	Cichlidae
GENUS	About 105
SPECIES	About 1,500

LENGTH
Up to 31½ in. (80 cm) in giant cichlid, *Boulengerochromis microlepis*

DISTINCTIVE FEATURES
Second set of (pharyngeal) jaws in throat; body usually deep and compressed. *Pterophyllum* (angelfish): disc-shaped body; high, sail-like fin. *Symphysodon* (discus fish): low fins. *Crenicichla* (pike cichlids): elongated body. Many species have strong coloration and markings.

DIET
Varies according to species; includes algae, crustaceans, mollusks, aquatic insects, other fish and detritus

BREEDING
Varies according to species; many species brood eggs and young in mouth

LIFE SPAN
Not known

HABITAT
Fresh and brackish waters; in South America most species found in rivers, in Africa greatest diversity of species occurs in lakes

DISTRIBUTION
Central and South America (including 1 species north to Texas); Caribbean islands; Africa; Madagascar; northwestern Middle East; India and Sri Lanka (2 species only)

STATUS
Many species abundant or common; 38 species critically endangered; 13 species endangered; 35 species vulnerable

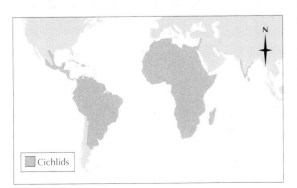

Cichlids

are laid, and fans the eggs continuously with fins and tail. Every few minutes the parents change over. The fanning probably increases the supply of oxygen for efficient development of the eggs, but it may also prevent fungal spores settling and germinating on them. At a later stage the eggs are ferried, a few at a time, from the spawning site to one of the pits dug in the sand. Each parent takes the eggs in the mouth to the pit, and as one fish makes a journey the other stands on guard. Later, by the same laborious process, the eggs are transferred to another of the pits.

Many cichlids hatch their eggs in their mouths, and for this reason are known as mouth-brooders. When the fry hatch, the parents keep a close watch on them. Usually, and especially in the early stages, the fry keep close together, but any that stray are taken in the mouth by one of the parents and returned to the fold.

Wealth of species in East Africa

A remarkable diversity of cichlids has arisen in the Rift Valley lakes of East Africa, and the great majority of these species live nowhere else. More than half of all known cichlids are found in just three of the lakes: Lake Victoria, Lake Malawi and Lake Tanganyika. There are 500 to 1,000 species in Lake Malawi alone.

The cichlid communities of East Africa provide a classic example of a phenomenon known as adaptive radiation. Cichlids have evolved to exploit every available food resource there, and most species are highly specialized. Among the many cichlid genera present in Lake Victoria, for example, are algal grazers (*Tilapia*), phytoplankton herbivores (*Enterochromis*), bottom-feeders (*Astatatilapia*), molluskivores (*Labrochromis*) and fish-eaters (*Prognathochromis*).

The Rio Grande cichlid, Herichthys cyanoguttatum, *is the only cichlid native to the United States.*

CIVET

The African civet is both common and widespread, but its acute senses and solitary, nocturnal habits help it avoid contact with humans.

THERE ARE 17 SPECIES of civets, although "civet" is a generic name not related to a scientific classification, and the grouping of these species varies according to which authority is followed. Originally derived from an Arabic term, the word civet was at first used to denote a scent obtained from the African civet, *Civettictis civetta*, and the Indian, or Oriental, civet, *Viverra zibetha*. Later it was also applied to the animals themselves, which are sometimes called civet cats. There is a separate article in this encyclopedia covering the related palm civets.

Civets are sufficiently diverse to make a general account of them problematic. Together with the mongooses and the genets, civets make up the Viverridae, a family of carnivores that shares characteristics of both the weasel and cat families. Like weasels, civets are long-bodied, with long, well-furred tails and sharp, pointed muzzles, but in habits and coat markings they resemble the small cats. The African and Indian civets remain the best known species. Among the others is the fanaloka, *Fossa fossa*, of Madagascar, also known as the Malagasy civet, which is 2 feet (60 cm) long. The tail accounts for one-third of

this length and is reddish gray in color, with rows of black spots. The otter civet, *Cynogale bennettii*, which is found from China south to Borneo, has a body 2 feet (60 cm) long, but with a tail of only 8 inches (20 cm). It has the habits of an otter and takes much the same food: fish and aquatic invertebrates.

Omnivorous diet

As a rule civets are solitary and nocturnal. They keep to dense undergrowth in forests or wooded savanna and rest by day, usually in the abandoned burrow of another mammal. Civets do not emerge to hunt until after nightfall. They can climb and swim well and capture some of their food in water. The adults' voice is a deep growl or a low-pitched cough.

The diet of civets is typically varied. It may include insects, crabs and other invertebrates, as well as frogs, lizards, snakes, fish, birds and mammals. Civets can kill prey up to the size of a hare or mongoose. They will also eat carrion and vegetable matter, such as fruits, roots and tubers. Civets readily eat foods that are either unpleasant or toxic to other animals, including

AFRICAN CIVET

CLASS	**Mammalia**
ORDER	**Carnivora**
FAMILY	**Viverridae**
GENUS AND SPECIES	***Civettictis civetta***

WEIGHT
15½–44 lb. (7–20 kg)

LENGTH
**Head and body: 27–35 in. (68–89 cm);
tail: 17–18 in. (44–46 cm)**

DISTINCTIVE FEATURES
**Low-slung, elongated body; pointed
muzzle; long, well-furred tail; ornate
pattern of spots and blotches on coat;
black eye patches and limbs; pale forehead;
crest of long hairs from head to tip of tail**

DIET
**Very varied, including rodents, birds,
lizards, snakes, frogs, crabs, insects,
carrion, fruits, roots and tubers**

BREEDING
**Age at first breeding: about 1 year;
breeding season: all year; number of
young: usually 2 or 3; gestation period:
60–72 days; breeding interval: 2 or 3
litters per year**

LIFE SPAN
**Up to 28 years in captivity, probably
much less in wild**

HABITAT
**Forest and savanna with abundance of
thick ground cover**

DISTRIBUTION
**Sub-Saharan Africa; absent from South
Africa (except Transvaal), most of Namibia
and driest regions of Ethiopia and Somalia**

STATUS
Common

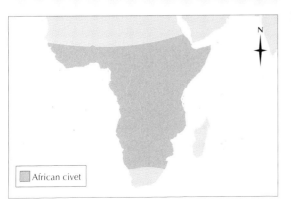

African civet

certain fruits, insects and millipedes. They are able to feed irregularly and if necessary can go without a substantial meal for up to 2 weeks.

Up to four young are born in a litter, which is usually located in a hole in the ground or among dense cover. The young develop quickly and are able to take solid food after 1 month. If they sense danger, young civets freeze until detected, whereupon they hiss and spit. Civets rarely breed in captivity.

Delicate perfume from rank odor

For centuries humans have collected musk from the pods, or glands, located near the reproductive organs in both sexes of the African and Indian civets. In the wild, civets smear this musk on landmarks within their home range, particularly on stumps at the intersection of paths, on rocks and on favorite fruit trees. Male civets augment these scented markings with dung piles and urine sprays. Civets do not have permanent dens to which they return; instead, they rest during the day in any convenient hollow or thick cover. The scent markings thus provide a vital means of orientation. The musk, a clear, yellowish or brownish complex of fats and essential oils, has the consistency of butter or honey. Offensive to the human nose in its concentrated form, the secretion is pleasant when diluted and accordingly has been used in some of the world's most renowned perfumes.

In Ethiopia particularly, and to a lesser extent elsewhere in Africa, civets are kept in captivity and the musk is removed from them several times a week. It is spooned out with a specially shaped wooden instrument, each animal yielding less than ⅛ ounce (3.5 g) per week. In 1934, 2½ tons (2.3 tonnes), worth about $120,000, were exported from Africa. The longevity of the industry is illustrated by the fact that King Solomon's perfume came from Ethiopia. However, synthetic products have now displaced natural musk in the perfumery trade.

Several species threatened

In spite of the long-standing practice of keeping civets in captivity, little is known of their life history and there have been few studies on the decline of civet species in the wild. While both the African and Indian civets are common within their respective ranges, other civets are rare. Overhunting and loss of habitat are regarded as leading contributory factors to the vulnerable state of many species, though road kills also account for a number of civet deaths annually. The I.U.C.N. (World Conservation Union) lists the Malabar large-spotted civet, *Viverra civettina*, a native of India, as critically endangered, and the Malagasy civet and otter civet as threatened.

CLAM

CLAMS BELONG TO THE group of mollusks known as bivalves or lamellibranchs. The term clam embraces a wide variety of species, including some that are only distantly related. Several edible species are commonly collected on mudflats along North American coastlines, among them soft-shell clams, quahogs (hard clams) and geoducks. Most clams are less than 20 inches (50 cm) long, but one species of giant clam, *Tridacna gigas*, reaches up to 50 inches (1.3 m) across and the whole mollusk may weigh 570 pounds (260 kg).

Open and shut case

A typical clam's body is enclosed in a shell made of two separate, hinged sections called valves. These are made of calcium carbonate (chalk) in a protein matrix, which is secreted by a structure known as the mantle, a flap of muscular tissue located immediately beneath both of the valves. The clam can shut its twin valves using muscles that run between them. When the muscles relax, the shell is pulled open by an elastic ligament at the hinge. A different muscle called the foot may be pushed out and used to creep along burrows or perform short jumps. Not all clams are mobile, however, and many spend all or much of their lives anchored by tough threads spun from a gland at the base of the foot. Some species, such as the Pacific geoduck of the west coast of North America, have a blade-shaped foot that enables them to dig down into the mud, where they are hidden from predators.

Dual purpose gills

On either side of a clam's foot, lying in the space enclosed by the mantle and shell, are the gills. These are used not only for breathing, but also to collect microscopic food particles from the water. The gills are typically large double flaps through which water is driven by the beating of cilia (minute hairlike projections) on their surfaces. In some clams water enters the mantle cavity at the front, passes through the gills and leaves at the rear. In many others water enters and leaves at the hind end through a pair of siphons, which are very long in a few species. When the food particles reach the gills, they become trapped in mucus and are carried to the mouth, which lies between a pair of ciliated lips. The inward water current created by the gills also supplies the mollusk with oxygen, while the outflow removes the animal's waste substances.

Giant clams

The shallows of coral reefs in the Indian Ocean and the South Pacific are home to the six species of giant clam, which have strongly corrugated valves that close rather like the teeth on a zipper. When a giant clam opens its twin valves a

*Most clams have hinged shells that protect the soft parts inside. This is **Tridacna crocea**, one of the six species of giant clam.*

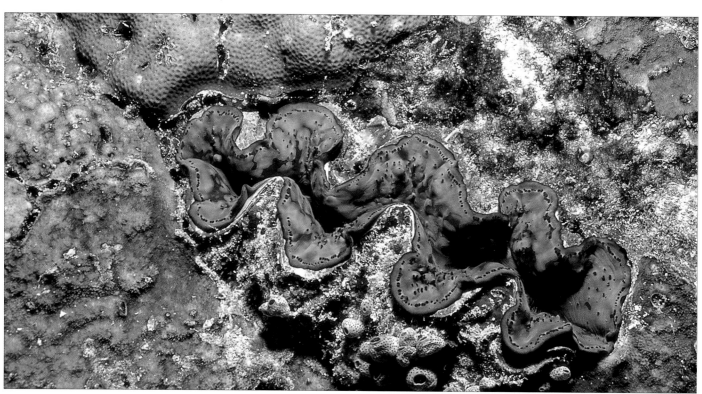

GIANT CLANS

CLASS	**Bivalvia**
SUBCLASS	**Lamellibranchia**
ORDER	**Eulamellibranchia**
FAMILY	**Tridacnidae**
GENUS	***Tridacna***
SPECIES	**6, including *T. gigas***

WEIGHT
***T. gigas*: up to 570 lb. (260 kg); much less in other species**

LENGTH
***T. gigas*: up to 4 ft. (1.2 m); other species usually 6–20 in. (15–50 cm)**

DISTINCTIVE FEATURES
Extremely large, corrugated shell; mantle (muscular tissue) visible through opening is bright purple, blue, green or yellow, depending on species

DIET
Filter phytoplankton from water; also feed on zooxanthellae (single-celled algae) growing inside their own shells

BREEDING
All species hermaphrodites, but no evidence of self-fertilization; age at first breeding: 3–4 years (male gonads), 5–6 years (female gonads); breeding season: according to lunar cycle; number of eggs: up to 1 billion

LIFE SPAN
Probably up to 50 years; longer in *T. gigas*

HABITAT
Coral reefs in shallow waters up to 115 ft. (35 m) deep

DISTRIBUTION
Tropical seas from Myanmar (Burma) south to northern Australia and east to Fiji

STATUS
Not known, but now scarce in many areas

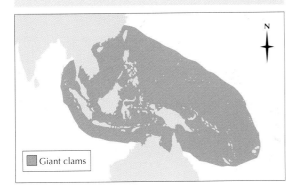
Giant clams

colorful mantle is revealed, which may be purple, blue, green or yellow. Large numbers of single-celled algae called zooxanthellae live in the outer fringes of the mantle, where they are exposed to the sun and can use its energy to make sugar by photosynthesis. The algae also receive some nutrients from the waste products of the host. In return the giant clam absorbs oxygen released by the algae and consumes a proportion of the algal growth as food.

Giant clams are edible and easy to harvest, but because they mature and reproduce slowly, they are vulnerable to overfishing.

Lunar breeders
Giant clams are hermaphrodites, that is they contain both male and female gonads. The male gonads normally mature first. There is no evidence, however, that giant clams are capable of self fertilization.

Each giant clam releases vast quantities of eggs and sperm into the water. If both male and female gonads have reached sexual maturity, the sperm are released first, the eggs about 1 hour later. Giant clams often reach a peak of spawning in the late afternoon during the second and fourth quarters of the lunar month. The planktonic larvae normally settle after about 10 days.

Killer myth
Although giant clams have a reputation for trapping the legs of divers in their shells, there are no authenticated reports of human fatalities caused in this way. Moreover, if someone were to tread on a giant clam, the shell would close slowly enough for the foot to be withdrawn.

CLAWED FROG

pale buff if placed against uniformly dark or light areas. A layer of mucus covers the clawed frog's body; its unpleasant smell may deter predators.

A mostly aquatic amphibian

The clawed frog is native to southern African streams, ponds and swamps; feral populations, descended from introduced animals, exist in North America and Britain. Unlike most other frogs, which spend a large part of their adult life on land, hiding in damp places and returning to water only to breed, the clawed frog lives in water most of the time, spending much of the time on the bottom. When it does venture on to land it is clumsy, but in water its strong hind legs and large webbed feet make it a powerful swimmer. The front legs are also used as paddles, and are not held beside the body as in typical frogs. To escape from predators, the clawed frog gives a violent thrust of the back legs, shooting itself backward or forward. The sharp claws may help the frog to grip boulders or plants in fast-running streams. The male may also use his claws to grasp the female's slippery skin during mating.

When many swamps and ponds dry up in summer, clawed frogs burrow into the mud at the bottom to remain cool and moist. If the mud also dries out, the frogs hop overland in search of permanent water. Otherwise the frogs leave their home waters and move overland only during heavy rains.

Breathing and feeding underwater

Although clawed frogs lead a primarily aquatic life, they still need to come to the surface to breathe. Their extra-large lungs enable them to gulp sufficient amounts of air to remain underwater for some time. A large frog surfaces every 10 minutes even when inactive, although it can survive longer than this if prevented from surfacing. With its very large areas of particularly permeable skin on the webs of its feet the clawed frog can absorb a high proportion of dissolved oxygen from the water.

Clawed frogs and their relatives do not have the extendable tongue that is used as a form of sticky weapon by other frogs. Instead they use

Huge webs of skin on the clawed frog's rear feet are richly supplied with blood vessels and help it to breathe extremely efficiently.

ALTHOUGH SOMETIMES REFERRED TO as a toad, the clawed frog belongs to the Aglossa, a group of tongueless frogs. It is larger than a common European frog, usually around 4 inches (10 cm) long, with large females reaching 5 inches (13 cm). The front legs are short and weak, each having four long, straight fingers. The back legs, however, are long and very muscular, with large webs between the toes. A clawed frog 3 inches (7.5 cm) long may have webbed feet 2 inches (5 cm) across. The South African name for the clawed frog, platanna, is derived from "plathander" meaning flat hands, a reference to these webs. The three inner toes on the frog's back feet have sharp black claws.

Clawed frogs can change color to match their environment. When the background is a contrasting mixture of light and shade, the frogs are mottled, but they can become almost black or a

AFRICAN CLAWED FROG

CLASS	**Amphibia**
ORDER	**Anura**
FAMILY	**Pipidae**
GENUS AND SPECIES	***Xenopus laevis***

ALTERNATIVE NAME
African clawed toad

LENGTH
Up to 5 in. (13 cm); female larger than male

DISTINCTIVE FEATURES
Flattened body; small forelegs with long, straight fingers; large, muscular hind legs with huge webbed feet; layer of mucus over body; color variable, changing according to environment

DIET
Freshwater invertebrates and small vertebrates; some carrion

BREEDING
Age at first breeding: 3–4 years; breeding season: rainy season; number of eggs: 500 to 2,000, laid singly; hatching period: 7 days; tadpoles metamorphose in 2 months; breeding interval: about 1 year

LIFE SPAN
Up to 15 years, longer in captivity

HABITAT
Freshwater streams, ponds and swamps

DISTRIBUTION
Southern Africa

STATUS
Locally common

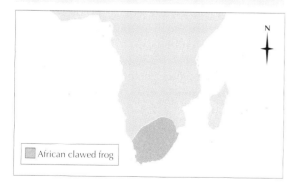

Filter-feeding tadpoles

At the end of the summer, when the ponds fill and the streams start to run again, clawed frogs begin to breed. During the breeding season, the males croak in the morning and in the evening. Eggs are fertilized after amplexus (the characteristic piggyback mating of frogs), and are laid singly on the stems of water plants or on stones. There may be 500 to 2,000 eggs. Each of them is 1 millimeter across and sticky, which enables it to adhere wherever the female places it.

The tadpoles hatch after about a week, depending on the water temperature, and for another week they hang motionless at the water's surface. During this time their mouths are closed and they live off the remains of the egg yolk, which has become enclosed in their body. Then they begin to feed on microscopic plants, but instead of scraping these off rocks and the surfaces of larger plants like most tadpoles, they suck water into their mouths and strain off the minute organisms. On each side of the mouth, just under the eyes, the tadpoles have a small tentacle that seems to be touch-sensitive. Experts believe that the tentacle helps keep the tadpole out of the mud as it feeds, hanging vertically downward near the bottom. The tadpoles take two months or more to turn into adult frogs, during which time they grow legs and lose their tail. They breed for the first time at 3–4 years.

The African clawed frog rarely leaves water, to travel over dry land. Its weak front legs make terrestrial movement slow and awkward.

their front feet in a fanning action to direct food in front of them. The food is then crammed into the mouth and pushed down the throat with the long, slender fingers on the front feet. Clawed frogs eat carrion, worms, crustaceans, aquatic larvae, small fish and even their own tadpoles.

465

CLICK BEETLE

Click beetles can jump up to 12 in. (30 cm) into the air, enduring a higher gravitational force than any other insect.

CLICK BEETLES ARE SMALL, hard-shelled, short-legged, elongated insects, generally about ¼ inch (7 mm) long. Most of the common species are black or dark brown, but some are red, yellow or green. The antennae may be simple or quite elaborately branched.

The harlequin click beetle, *Chalcolepidius zonatus*, is a 2-inch (5 cm) long insect with black and white stripes and is usually found on fallen trees in South American forests. The elegant click beetle, *Semiotus affinis*, is a large and spectacularly colored species, which inhabits the rain forests of the American tropics.

Jumping beetles

Most click beetles are active at night and hide away in the daytime. Apart from the luminosity of some tropical species, their most notable feature is the ability to jump with an audible click when turned on their backs.

In click beetles the first and second sections of the thorax are hinged, and on the underside of the first there is a spine directed backward, the tip of which rests just over the edge of a cavity in the second. The spine is pressed against this edge, and as the hinge between the two sections moves, it causes the spine to slide until its tip passes over the edge and snaps into the cavity

with enough force to jar the whole body of the insect and throw it into the air. This mechanism is peculiar to the family Elateridae.

If a click beetle is put on its back it can be seen to arch its body using its thoracic hinge just before jumping. The beetle cannot be sure of landing the right way up at the first jump, and in fact often fails to do so, but by repeated jumps it eventually lands on its feet.

The legs of a click beetle are so short that they cannot be used to right the beetle when it is inverted, so the click mechanism must be of considerable value for this reason alone. However, scientists believe that the mechanism may serve another purpose. If caught and squeezed, the beetle always clicks repeatedly. This may be no more than a response to being held off balance but it is quite likely that a young, inexperienced bird or lizard, finding and seizing a click beetle, might be so startled by the strength of the clicks that it will drop its prey, which then has the opportunity for escape.

In common with other insects, click beetles employ another defense mechanism, that of thanatosis, or shamming death. If it is touched or senses danger, a click beetle draws in its legs and lets itself fall to the ground until the danger is past. However, the effectiveness of this act is uncertain. Some predators, such as toads, will take only active creatures, but not all insect-eaters limit themselves to prey that is alive and moving.

Major crop pests

Adult click beetles feed on leaves, mostly at night, and are also attracted to sweet liquids. Some harmful species are trapped by putting out sweet baits to attract them. The larvae of some click beetles live in rotting wood, but those of the most abundant species (*Agriotes* and *Athous*), known as wireworms, feed on the roots, bulbs

---

CLICK BEETLES

PHYLUM	**Arthropoda**
CLASS	**Insecta**
ORDER	**Coleoptera**
FAMILY	**Elateridae**
GENUS	**Many, including *Agriotes*, *Athous*, *Chalcolepidius* and *Semiotus***
SPECIES	**About 8,000, including striped elaterid beetle, *Agriotes lineatus* (detailed below)**

ALTERNATIVE NAMES
Skipjack; wireworm (larva only)

LENGTH
Adult: ¼–⅓ in. (7–8 mm)

DISTINCTIVE FEATURES
Adult: elongated body with very hard outer surface; large head; short legs; brownish overall but often covered with grayish microscopic hairs

DIET
Adult: leaves; flower pollen and nectar. Larva: roots and seeds.

BREEDING
Age at first breeding: up to 7 years; breeding season: eggs laid May–June; number of eggs: 150 to 200; hatching period: 25–60 days; breeding interval: 1 generation every 6 years

LIFE SPAN
2–7 years

HABITAT
Most habitats; larva in wood or soil

DISTRIBUTION
Worldwide

STATUS
Abundant

---

and tubers of plants and also on seeds lying in the ground before germination. The larvae are equipped with a hardened cuticle which is resistant to crushing. Wireworms are elongated, cylindrical, tough-skinned larvae, usually yellow in color, and are among the most serious of all insect pests. They attack most cultivated crops, but especially cereals and potatoes.

The eggs are laid in the soil and the larvae develop slowly, taking 2–6 years to reach full size. In the commonest British wireworm, *Agriotes obscurus,* the larval life duration is estimated at 5 years. Only a few weeks are spent as a pupa, but the adult beetles may live for 10 months or a year, overwintering in the soil in temperate and cold climates.

Agriotes and *Athous* wireworms are serious agricultural pests. In land that is already cultivated, some measure of control can be achieved by getting rid of weeds (which support the wireworms between crops), frequent working of the soil and careful use of insecticides. However, of all insects harmful to agriculture, wireworms are among the most difficult to destroy.

When grassland is ploughed it is usual for an economic entomologist to sample the soil and determine the approximate number of wireworms present per acre of the land. If the figure exceeds 1 million, no crop will have a very good chance of success. A figure of over 600,000 wireworms is a serious threat to most crops on light soil, but on heavy soil peas can be planted and also barley, the least susceptible of the cereals. For all crops except potatoes, a figure of 300,000 to 600,000 wireworms per acre is regarded as an acceptable population.

Click beetle larvae can cause massive damage to crops, but some species (such as Ampedus cardinalis, *pupa, below) feed on rotten wood and are less harmful.*

CLIMBING PERCH

Only one species of climbing perch, native to southern Asia, can travel over dry land. However, the African climbing perch of the genus Ctenopoma *(above) share their more famous relative's ability to live in polluted water.*

ALTHOUGH LONG FAMILIAR to scientists under the name *Anabas scandens*, the climbing perch is now generally referred to as *Anabas testudineus*. It grows up to 10 inches (25 cm) long, and is gray-green to grayish silver in color with brown fins. The fish has a dark patch behind each gill cover and another at the base of the tailfin. The climbing perch is found across southern and Southeast Asia, from India and Sri Lanka to southern China, the Philippines and the Malay Archipelago. A number of other fish are also called climbing perch, including at least 22 African species in the genus *Ctenopoma* and two South African species in *Sandelia*.

A fish out of water

The climbing perch's most distinctive characteristic is its ability to travel overland to find a fresh pond when its home waters have dried up. The perch does this using its gill covers, pectoral fins and tail. The gill covers have spines on their hind margins, which the fish digs into the ground with a side-to-side rocking motion. The pectoral fins are used as props to help the forward thrust from the tail. The action is half a wriggle and half a seal-like lollop and enables the fish to travel at about 200 yards (180 m) per hour. Although the overland speed of the climbing perch is slow, the fish can stay out of water for a long time. The people of India and Malaysia carry them for days on end in moistened clay pots, ensuring a supply of fresh fish.

The climbing perch's ability to breathe out of water is possible because of its large gill cavity, divided into two compartments. The smaller and lower of these contains the normal gills. The larger upper part, called the labyrinth, contains a rosette of concentrically arranged plates with wavy edges, the whole richly supplied with a network of fine blood vessels. In fact, the rosette works like a piece of lung. Air is swallowed and passes by an opening on either side of the throat into the rosette chambers, the opening being controlled by a valve. The spent air leaves through the gills' exit.

The climbing perch does not only use this breathing apparatus when it is out of water. Like the true lungfish, and other fish that have lungs or lung-like organs, the climbing perch can survive in water that is polluted or foul from rotting vegetation and which is therefore low in oxygen. In such water it rises to the surface to gulp air. The climbing perch is one of several fish which drown if held underwater.

At the beginning of the dry season the climbing perch burrows into mud and enters a resting stage, like the well-known lungfish. At other times, and usually in the early morning or during a rainstorm, the fish travel over the ground in troops. Climbing perch are tolerant of sudden temperature changes, and feeding offers few problems since they take a wide variety of animal and plant food.

Male parental care

The male climbing perch turns the female onto her back in order to copulate. After several false pairings a few eggs are laid, which usually float up into the nest. Spawning involves a great

CLIMBING PERCH

CLASS	**Osteichthyes**
ORDER	**Perciformes**
FAMILY	**Anabantidae**
GENUS	***Anabas, Ctenopoma*** and ***Sandelia***
SPECIES	**More than 25, including *A. testudineus* (detailed below)**

ALTERNATIVE NAMES
Climbing fish; climbing gourami; labyrinth fish

LENGTH
Up to 10 in. (25 cm)

DISTINCTIVE FEATURES
Perchlike form; body elongated and moderately deep; overall color gray green to gray silver, occasionally very dark; fins translucent or brown, less frequently yellowish; dark blotch on both gill cover and tail; dark band from eye to mouth

DIET
Very varied, including algae, aquatic plants and small animals

BREEDING
Number of eggs: up to 1,500; hatching period: 24–36 hours

LIFE SPAN
Not known

HABITAT
Fresh water, including weedy rivers, streams, ponds, irrigation ditches and flooded ricefields; also foul-water sites such as drains

DISTRIBUTION
India and Sri Lanka east to southern China and south to Philippines and Malay Archipelago

STATUS
Not known, but probably not threatened

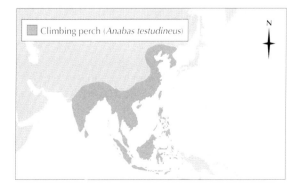

Climbing perch (*Anabas testudineus*)

The climbing perch (Anabas testudineus, above) thrives in temporary wetlands, moving overland in search of a new home when its home waters dry up.

number of such actions. The eggs' buoyancy comes from oil droplets contained within them. Eggs that fall to the bottom are picked up by the male and spat into the nest. When spawning is complete, the male drives the female away from the nest, sometimes biting her, and undertakes the care of the brood alone.

It is surprising, in view of their omnivorous diet, that climbing perch do not eat some of their young, as so many freshwater fish are prone to do, after nursing them through early infancy. From the time the eggs float to the surface, neither of the parents show any further interest in them. The eggs hatch in 2–3 days.

End of a legend

The western world first learned of the existence of the climbing perch in 1797, when Lieutenant Daldorff of the Danish East India Company, stationed at Tranquebar, published a description of the species. He recounted a local legend that the fish climbed palm trees and sucked their juices. Daldorff confirmed the story by saying he himself had found such a fish in a slit in the bark of a palm growing near a pond.

The fish was known as a climbing perch for the next 250 years. It is true that the species will climb the trunk of a leaning tree, as will the marine fish known as mudskippers. Moreover, the climbing perch is sometimes found well up a tree, in a crotch, fork or crevice in the bark.

However, in 1927 an Indian expert on fish, Dr B. K. Das, discovered that climbing perch found in trees do not get there by their own efforts. When troops of perch are traveling in search of fresh ponds, crows and kites may swoop and carry some of them off. The birds leave their catch in various places, including trees, where the fish can live for several days.

CLOTHES MOTH

Adult clothes moths are harmless: it is the larvae (above) that cause extensive damage by feeding on household fabrics.

TWO SPECIES OF SMALL MOTHS are known as clothes moths. These are the common clothes moth, *Tineola bisselliella*, and the casemaking clothes moth, *Tuea pellionella*. Both species belong to the family Tineidae. A third species, the tapestry moth, *Trichophaga tapetzella*, may occasionally be found in clothing, though it is not technically a clothes moth.

The larvae of clothes moths damage furs and woollen fabrics by feeding on them; the adult moths, sometimes called millers, are harmless to materials. Clothes moth larvae, however, have a complicated digestive system and are among the very few animals capable of digesting keratin, a protein present in wool, hair, horns, fur, feathers and hoofs. They are particularly attracted to articles which are soiled with substances such as oil from hair, sweat and urine. However, the larvae cannot digest nonprotein-based materials, such as synthetic fibers.

Also known as the webbing clothes moth, the common clothes moth is approximately ¼ inch (6.5 mm) long and has a wingspan of ½–⅔ inch (12.75–16.75 mm); females are larger than males. The forewings (the part seen when the moth is at rest) are pale buff or golden, plain and glossy or shiny. The larva is white with a brown head and reaches a length of ⅓–½ inch (8.5–12.75 mm), depending on the environment. The larva builds a cocoon made out of fibers and its own silk in which it pupates.

The casemaking clothes moth is similar in appearance to the common clothes moth. It has three dark spots on each wing and a brownish tinge, though it may not appear as shiny as the common clothes moth. The larva is more easily distinguished by its habit of making a case of silk and fiber that it lives inside, feeding from either end and retreating into the case when disturbed. The pupa is formed inside the case, which then serves as a cocoon.

Thrive in the dark

Clothes moths dislike both sunlight and artificial light, and are rarely seen. When the items on which they are resting are disturbed, clothes moths either seek cover or fly to a darker area to conceal themselves. They can squeeze through narrow crevices and find their way easily into cupboards or chests of drawers. Adult moths

COMMON CLOTHES MOTH

PHYLUM	**Arthropoda**
CLASS	**Insecta**
ORDER	**Lepidoptera**
FAMILY	**Tineidae**
GENUS AND SPECIES	***Tineola bisselliella***

ALTERNATIVE NAMES
Webbing clothes moth; miller (adult only)

LENGTH
**Body: about ¼ in. (6.5 mm);
wingspan: ½–⅔ in. (12.75–16.75 mm)**

DISTINCTIVE FEATURES
Adult: buff or golden-colored wings and body; wings fringed and unspotted; head sometimes appears reddish. Larva: white body and pale brown head.

DIET
Adult: does not feed. Larva: wide range of foodstuffs, including natural fabrics (woollen clothing, stored wool and rugs), furs, upholstered furniture, animal bristles in brushes, fish meal, flour, fungal material and carrion.

BREEDING
Age at first breeding: 4–6 months; breeding season: all year; number of eggs: up to 200; hatching period: 3–21 days; breeding interval: about 2 generations per year, depending on environmental conditions

LIFE SPAN
Adult male: up to 28 days; adult female: dies shortly after laying eggs

HABITAT
Largely dependent on stored natural fabrics inside buildings but originally associated with nests of small mammals, birds, social bees, social wasps and social ants

DISTRIBUTION
Virtually worldwide

STATUS
Globally abundant

seen flying in the home are probably males, as females usually hide in the folds of clothing and mostly travel by running or hopping.

Clothes moth larvae can live on clean wool and fur, but much prefer clothing and fibers left undisturbed for a long time, such as stored fabrics, areas of carpeting covered by furniture and the crevices in upholstered furniture. They bite through and scatter far more fibers than they eat, which accounts for the great amount of damage caused by these insects.

Growth of the larvae

The female common clothes moth lays up to 200 eggs over a period of 2–3 weeks. These will take 3–21 days to hatch, depending on the environment; clothes moths develop best in a heated, dark room with about 75 percent relative humidity. The larvae's growth rate also varies widely with the availability of food. Clothes moth larvae cause damage to natural materials rather than artificial substances, and on raw wool or rabbit fur they may become adults in 3 or 4 months. Adult females die shortly after laying their eggs, whereas adult males may live for up to 28 days.

Much is known about the breeding of the casemaking clothes moth and there are many breeding similarities between this species and the common clothes moth. The lifestyle of the larvae differs slightly and adult casemaking clothes moths tend to be more active fliers than adult common clothes moths.

Control of clothes moths

Clothes moths have several natural predators and parasites, which can be used to control serious infestations. The larvae of the window fly, *Scenopinus fenestralis*, prey on clothes moth larvae and may be found, among other places, in household carpets. A small species of parasitic chalcid wasp, *Spathius exarator*, lays its eggs inside the bodies of clothes moth larvae.

Clothes moths are adversely affected by mothproofing, drycleaning, synthetic fabrics and improved standards of hygiene in homes. If woollens are kept scrupulously clean and carpets regularly cleaned with a sweeper or vacuum cleaner it is unlikely that clothes moths will infest them. Sudden changes of temperature also kill the insects, hence the value of putting furs in cold storage for the summer.

Life in the wild

Clothes moths do not just rely on human clothing and fabrics for food. In their natural state the species live in bird, mammal and insect nests and on corpses. Here they have a diverse diet, including a mixture of detritus and animal fur or wool, along with a certain amount of fungal material.

Clothes moth damage has reduced in recent years as a result of improved hygiene and more frequent drycleaning, but the cost of controlling these insects and repairing the damage they cause is still considerable. The species are plentiful in their natural environment and are likely to thrive for many years to come.

COALFISH

Primarily an inshore species, the coalfish is usually encountered in midwater near rocky coasts.

A MEMBER OF THE cod family, up to 3½ feet (1.1 m) long and characterized by the dark color of its back, the coalfish was given its name in about the year 1600 because of its black coloring. About 30 years later it also became known as the saithe. In North America the species is known as the pollack.

Coalfish occur in much of the North Atlantic. The adults cover considerable distances, particularly when making seasonal migrations to offshore spawning grounds in the winter, and the fry are carried hundreds of miles by currents. Thus, although the species' spawning grounds are restricted, mature coalfish are widely distributed across the Atlantic from the eastern seaboard of North America to northwestern Europe.

Predatory shoals

The members of the cod family are voracious feeders, smaller fish being an important part of their diet. Indeed, coalfish feed on the fry of a near relative, the Atlantic cod. They first surround a shoal of fry on all sides, driving the small fish into a dense mass. Then, by a sudden maneuver, they force the fry upward towards the surface, attacking from below while seabirds gather overhead to attack the fry from above.

The coalfish feeds on other, smaller fish as well as on small crustaceans, such as amphipods. As with so many predatory fish it swallows its food whole and takes 5–6 days to digest a fish meal, the bones becoming softened toward the end of this time. A crab swallowed whole will be digested in a similar period but its exoskeleton becomes softened first. Smaller crustaceans, such as amphipods, are digested in 3–3½ days.

Nurseries on the high seas

The main spawning grounds of the coalfish are off the western coasts of the British Isles, in the North Sea and off Norway. The eggs are laid from January to May, each egg being 1 millimeter in diameter and surrounded by a transparent membrane within which is a tiny globule of oil that gives the egg buoyancy. Although the eggs are laid at depths of approximately 330–660 feet (100–200 m) they quickly rise to the surface and float there. Inside the egg membrane, and forming part of the egg, is a supply of yolk on which the embryo feeds. Each of the newly hatched fry carries a yolk sac on its underside, the contents of which nourish it during the early days of life. The yolk sac is larger than the young coalfish and more buoyant, so the fry float

COALFISH

CLASS	**Osteichthyes**
ORDER	**Gadiformes**
FAMILY	**Gadidae**
GENUS AND SPECIES	***Pollachius virens***

ALTERNATIVE NAMES
Saithe; coley; billet; glassan; black pollack (U.S. only)

LENGTH
Up to 43 in. (1.1 m)

DISTINCTIVE FEATURES
Adult: body black, greenish brown or olive, lighter on sides and with silver-gray belly; 3 dorsal fins and 2 anal fins; small pelvic fins well forward on body; notched caudal fin; creamy-white, nearly straight lateral line. Young: minute chin barbel.

DIET
Adult: mainly other fish, in particular young cod, capelin and sand eels; also crustaceans. Young: crustaceans and small fish.

BREEDING
Age at first breeding: 2–3 years; breeding season: January–May; number of eggs: average 250,000; hatching period: 6–9 days; breeding interval: 1 year

LIFE SPAN
Up to 25 years

HABITAT
Offshore and inshore waters, often associated with rocky coasts and seabeds

DISTRIBUTION
North Atlantic from eastern North America north to southern Greenland and east to Scandinavia; Baltic Sea; White Sea (Arctic waters to north of European Russia)

STATUS
Common in some areas; size of populations unknown

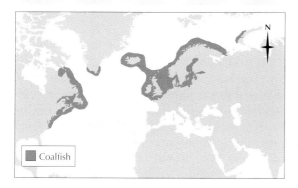

Coalfish

upside down at the surface. As the yolk becomes used and the yolk sac shrinks the fish slowly turns the right way up. It is finally ready, 6–9 days after hatching, to swim deeper and to start feeding on very small crustaceans.

Internal buoyancy control

In common with other members of the cod family the coalfish has a swim bladder. This is an elongated silvery bag readily seen when the body of the fish is opened up. It lies just under the backbone among the organs of the fish's digestive system. A substance known as isinglass is made from this maw, or sound as it is also called. This is a kind of gelatine once used for preserving eggs, clarifying beers and wines and for a variety of other purposes. The walls of the swim bladder are richly supplied with blood

vessels and in marine fish these are concentrated to form a gas gland capable of giving out a mixture of oxygen and nitrogen in roughly the same proportions as in air. Another knot of blood vessels, known as the oval, absorbs these gases. Using these a fish can inflate or deflate the bladder, adjusting the volume to keep its body density as a whole equal to that of the surrounding water. As the fish goes deeper and the pressure on its body increases, the swim bladder is inflated to compensate. The reverse process takes place when the fish rises.

Marine fish that possess a swim bladder float and rest by adjusting it according to the depth in much the same way that submarine buoyancy tanks operate. Fish lacking a swim bladder must constantly keep swimming to avoid sinking.

Coalfish are voracious predators and feed readily on smaller fish, including other members of the cod family.

COAST

Far more species live in the coastal biome than out to sea. Tides and rivers bring a never-ending supply of vital nutrients on which, ultimately, all life there depends.

OF ALL THE BIOMES on Earth, the coastal biome is one of the most varied. It contains a wealth of different habitats that are home to hundreds of thousands of plant and animal species. A much wider variety of species inhabits the coastal biome than the deep water of the open sea, known as the oceanic biome. The majority of the species found in the coastal biome are permanent residents, but many others travel there from deeper water and from dry land, to feed and breed.

Zones of life

Most zoologists divide the coastal biome into two major zones according to water depth and distance from land. The strip of shallow water along coastlines is known as the neritic zone, and supports the greatest abundance of marine life. In this zone different types of seabed, including gravel, sand, mud and coral reef, create different habitats, each of which has characteristic species. The shallows of the neritic zone are the submerged edges of continental landmasses and rarely exceed depths of 650 feet (200 m).

Farther toward land is the littoral, or intertidal, zone, which is covered and uncovered every day by the tides. All of the organisms that live here must be able to resist the relentless action of the tides. They achieve this by burying themselves in the seabed, by seeking refuge among rocks, corals and underwater plants, or by securing themselves to immobile objects. The littoral zone contains many habitats worldwide, including rocky shores, mudflats, sandy beaches, salt marshes and mangrove swamps. On the landward side these habitats give way to a range of strictly terrestrial habitats, such as sand dunes, cliffs, brackish (slightly salty) lagoons, scrub and freshwater marshes.

Essential tiny plants

The coastal biome teems with life because its shallow waters receive plenty of light and are constantly replenished with large quantities of vital nutrients. A host of bacteria and other microscopic organisms, collectively known as detritivores, lives on the seabed throughout the world's seas and oceans. These tiny organisms feed on dead plant and animal matter, in the process releasing nutrients into the water. Chief among these nutrients are mineral salts, nitrogen, phosphorus and iron. Upwelling currents carry these nutrients to the surface of the sea and on to the shallow waters of the neritic zone. The action of tides and rivers brings yet more nutrients to coastal waters by washing silt away from littoral and terrestrial habitats. Coastal shallows therefore maintain high nutrient concentrations.

Single-celled plants called phytoplankton grow in vast quantities in the coastal biome. They use the combination of water, carbon dioxide, maximum sunlight and plentiful nutrients to photosynthesize food. Phytoplankton form the base of most food webs in the seas and oceans, and the enormous numbers of these plants in coastal waters is the reason why so many other plants and animals thrive there.

Food webs

In coastal food webs the stage above phytoplankton is occupied by zooplankton, a collective term for a tremendous diversity of organisms. Zooplankton include thousands of tiny invertebrate species, such as krill (shrimplike animals), copepods (a type of crustacean) and sea butterflies (marine snails), as well as the larvae or young of larger animals, particularly fish and crustaceans. Zooplankton are too small to swim against currents and tides and cannot manufacture their own food by photosynthesis. Instead they drift along, grazing the blooms of phytoplankton or feeding on the other members of the zooplankton.

In turn zooplankton provide food for nekton, a term encompassing any marine animal that can swim freely. Small nekton comprise small fish, squid and jellyfish. Larger nekton include big fish, such as sharks, tuna and mackerel, turtles and large mammals, including seals, dolphins, porpoises and whales. Also included in this stage of coastal food webs are diving seabirds, such as penguins, auks, terns and boobies.

All phytoplankton and zooplankton and most species of nekton live in the water column, and are collectively known as pelagic species. When pelagic species die they sink to the seabed, where they provide food for bottom-dwelling, or benthic, species. The nutrient cycle is completed by the benthic species because they digest the bodies of pelagic species, returning the nutrients to the water and thus eventually to the phytoplankton at the sea's surface. Benthic species include seaweed, sponges, worms, echinoderms (such as starfish, basket stars, sea cucumbers and sea urchins), crabs, mollusks (such as clams, barnacles and mussels) and bottom-feeding fish (such as flounders, soles and stingrays).

Rocky shores

Coastlines strewn with rocks and boulders are a harsh environment, and the extent to which plants and animals are tolerant of the demanding physical conditions found there determines which part of this habitat they occupy. The dominant environmental factors of rocky shores are the degree of immersion in salty water, the strength of the wave action and the temperature differential between the water and exposed sunny rocks.

Rocky shores are often divided by scientists into bands roughly parallel to the water's edge. In the highest band, which is covered only during the highest tides and by spray, lichens and blue-green algae grow on the relatively dry rocks. Seabirds use the rocks as perches and roosting sites. Lower down the rocks lies a band that is covered and uncovered daily by the tides.

Predators such as the harbor seal, Phoca vitulina, *are drawn to coasts by the rich harvest of fish and invertebrates.*

To survive in the middle band of rocky shores animals must be tough enough to survive daily batterings by the tides and alternate periods spent underwater and on dry land.

Mollusks, sea anemones and sea firs are common in this band because they strongly adhere to the rocks and can resist powerful waves. The calmer, lowest band of rocky shores is only rarely exposed to the air during very low tides. It is dominated by kelp and other species of seaweed, which are close to the light and yet protected from drying out. Seaweed provides shelter for a wide variety of animals, including starfish, sea urchins, small fish and octopuses. This community of animals in turn attracts visiting predators, such as larger fish, otters, seals and sea lions.

Sandy shores and mudflats

Although rocky shores are pummeled by waves the tidal disturbance is often greater on sandy shores, where the ground literally moves with every tide. Unlike seashore rocks, beaches do not have life teeming on their surface. This is because the ability to burrow into the sand is vital for survival there. It follows that seaweed is absent from sandy shores; the only plants to be found are microscopic algae called diatoms.

The chief food source on sandy shores is the dead organic matter brought by each tide. This is decomposed by bacteria, nematode worms and copepods, which are preyed on by polychaete worms, such as ragworms. The size of the sand grains and the slope of the beach determines which animals are present. For example, the combination of small grains and a shallow slope results in more water being retained within the beach, which makes burrowing much easier but reduces the amount of available oxygen, particularly at over 1 inch (2.5 cm) below the surface.

Sandy shores often support a superabundance of specially adapted animals, such as sand fleas, sand dollars, sea mice, tube worms and cockles. However, the variety of species present is limited because relatively few species can survive in the thin available habitable layer. If the slope of a beach is very shallow, organic particles settle out to form mudflats. A similarly low diversity of species is found in mudflats, but they are present in even greater numbers because the mud is richer in nutrients, the tide being too slow to wash them out again.

In winter mudflats are the main feeding grounds for many species of wading bird, including various sandpipers, oystercatchers, plovers, godwits and curlews. The topmost sections of beaches and mudflats are used by seals for resting and breeding, while tidelines attract a range of scavengers in search of washed up scraps, among them gulls, birds of prey, crows, foxes and even bears.

Salt marshes and mangrove swamps

On the more sheltered coasts that experience minimal wave action, the exposed mud is colonized by larger plants. Aided by secretions from diatom algae, these plants bind the mud together. The mud is deficient in oxygen, but the specialized plants have evolved a tangled network of aerial roots; at low tide the formerly submerged roots are exposed and can absorb oxygen from the air. In temperate regions this process forms a habitat known as salt marsh, which is dominated by a small number of salt-tolerant plant species, particularly cord grasses (in the genus *Spartina*), rushes (*Juncus*) and plantains (*Plantogo*). In tropical regions mangrove trees, rather then grasses, dominate.

Both salt marshes and mangrove swamps are a natural form of coastal defense. They play a major part in the creation of new land by gradually colonizing the mudflats that lie on the seaward side. These habitats also support a rich variety of wildlife. Salt marsh is particularly important for ducks and geese. Mangrove trees shelter large breeding colonies of cormorants, ibises, egrets and herons, and are home to a small selection of resident mammals and reptiles.

Conservation

The coastal biome faces growing threats from human activity in many parts of the world. Yet it is of incalculable importance to humans as well as wildlife. For example, coastal waters account for 90 percent of the total production of the world's commercial fisheries; in Asia, fish and seafood provide over a billion people with their primary source of animal protein. Coasts are also of vital importance for industry and tourism. Urban areas and tourist resorts are often concentrated along coasts, placing increasing pressure on fragile coastal habitats.

Marine pollution is one of the most serious general threats to the coastal biome. The pollutants include raw sewage, refuse, oil spills and industrial and agricultural chemicals, which are washed into the sea by rivers. The biome also faces many specific threats; beaches are damaged by tourist developments, coastal waters are overfished and salt marshes and mangrove swamps are destroyed by flood protection and land reclamation measures. For instance, 30 square miles (80 sq km) of coastal wetland per year are destroyed in the United States alone and yet 75 percent of American wildfowl need coastal wetlands to breed.

The coastal biome is economically important, but the commercial exploitation of coasts must be balanced against the needs of the many plants and animals found there.

Mangrove trees are raised above the coastal mud on a mass of exposed roots, some of which have small openings through which air can enter.

COATI

The coati (N. narica, above) is also known as the coatimundi, a name first used by Central and South American peoples to refer to a lone male. The word "coati" was applied to the animal only when it was encountered in groups.

THE COATIS ARE FOUR small, carnivorous mammals related to the raccoon, red panda and ringtail, or cacomistle. They range in size from 15 inches (38 cm) in the mountain coati, *Nasuella olivacea*, to 16–26 inches (41–67 cm) long in the three species *Nasua nasua*, *N. nelsoni* and *N. narica*. The ears are small and the forehead is flat, running down to a long, mobile snout. The black nose at the tip is moist, which helps give coatis an excellent sense of smell. The coat color is reddish brown to grayish brown, with yellowish underparts and black and gray markings on the face. The tail is up to 28 inches (70 cm) long and is banded.

Both widespread and localized

The four species of coati inhabit the forests of South and Central America. One species, the ring-tailed coati, *N. nasua*, is also found in Arizona, New Mexico and Texas, where it is presently extending its range northward. The endangered island coati, *N. nelsoni*, is found only on Cozumel Island, which lies to the northeast of the Yucatan Peninsula, Mexico.

Bands of females and young

Young and female coatis live in bands of up to 20, whereas males lead a solitary life, except during the breeding season. Each band of coatis has a home range, the borders of which overlap with those of other bands, but groups rarely meet because they tend to stay near the centers of their ranges. When separate bands do meet, there is usually confrontational behavior, although this rarely develops into fighting.

The members of a coati band forage together, retiring just after sunset to a favored tree, where they remain until sunrise. The coatis sleep curled up with their tails over their faces, in nests of twigs and creepers placed in a fork or on a mat of branches. Several coatis may share a nest. There is no fixed social hierarchy and little aggression between band members.

Although coatis are very active throughout the day, there are periods of rest, during which the coatis groom themselves to remove parasites. Mutual grooming probably also strengthens the bonds between the individuals in a group. While a band of coatis is out on a foraging expedition, the animals may become separated, but before long the small parties search for one another, making chittering calls to attract attention.

Foraging in the forest

Coatis adopt two gaits when climbing, either ascending paw-over-paw or galloping up wide trunks with forefeet and hind feet clutching the bark together. To descend, they move headfirst with their hind feet held backward, rather like squirrels. Coatis will jump from branch to branch, but not over any great distance, and seem to fall quite often.

Coatis forage both in the leaf litter on the forest floor and up in the trees. On the ground they run about, sniffing to locate small animals hidden underground and in rotten logs. In this way many kinds of invertebrates can be snapped up, including millipedes, earthworms, termites, snails and tarantulas. Coatis catch mice and lizards after flushing them out of their holes, and also prey on land crabs and caecilians.

One trait typical of a hunting coati is persistance. Coatis have been seen to spend half an hour digging out lizards, burrowing so far down

COATIS

CLASS	**Mammalia**
ORDER	**Carnivora**
FAMILY	**Procyonidae**
GENUS AND SPECIES	**Ring-tailed coati,** *Nasua nasua*; **white-nosed coati,** *N. narica*; **island coati,** *N. nelsoni*; **mountain coati,** *Nasuella olivacea*

ALTERNATIVE NAME
Coatimundi (*N. nasua, N. narica*)

LENGTH
Head and body: (*N. nasua, N. narica, N. nelsoni*) 16–26 in. (41–67 cm); (*Nasuella olivacea*) 15 in. (38 cm)

DISTINCTIVE FEATURES
Long, slender, highly mobile snout; tail very long and bushy with black rings

DIET
Mainly invertebrates, rodents and fruits; also crabs, amphibians, lizards and reptile eggs

BREEDING
Age at first breeding: 2–3 years; breeding season: usually January–March; number of young: 2 to 7; gestation period: 70–80 days; breeding interval: probably 1 year

LIFE SPAN
Up to 18 years in captivity

HABITAT
Woodland (*N. nasua, N. narica, N. nelsoni*); montane cloud forest (*Nasuella olivacea*)

DISTRIBUTION
Southern U.S. south to Argentina (*N. nasua, N. narica*); Cozumel Island, off northeastern Yucatan Peninsula, Mexico (*N. nelsoni*); Andes Mountains in western Venezuela, Colombia and Ecuador (*Nasuella olivacea*)

STATUS
Common: *N. nasua, N. narica*; endangered: *N. nelsoni*; rare: *Nasuella olivacea*

Coatis

Shortly before she is due to give birth the female coati (N. nasua, left) leaves the rest of her group and builds a treetop nest ready for her litter.

that their bodies disappear from sight. Land crabs are dealt with by deftly flicking them into the open and ripping off their claws. If the prey escapes, the entire band of coatis may set off in pursuit. Prey is killed either by being bitten in the neck, in the case of lizards and rodents, or by being rolled under the front paws until mangled. Coatis may use this method to crush snail shells and to remove insect hairs and stings.

Coatis also eat a variety of fruits, including wild bananas, figs, mangoes and papayas, either feeding on fallen fruit or climbing the trees to pluck it. Sometimes a band of coatis splits up and one part waits under the trees to pick up fruit dropped by the other, which later returns to take its share of the food.

Young reared in nests

Coatis usually mate during the dry season. At this time the bands of females and juveniles are joined by males, which are aggressive toward one another, each male fiercely guarding his own band; rival males are also discouraged by the females of each band. In coatis the gestation period is 10–11 weeks, and just before the litter is due the pregnant female leaves her band to make a nest in a tree. The young, usually two to seven in number, are born in the nest and stay there for 5 weeks, by which time they can run and climb well enough to keep up with their mother when she rejoins the band. They will continue to associate with her, while gradually becoming more independent. Young males leave and adopt a solitary life when sexually mature, at 2–3 years.

COBRA

Cobras (Indian cobra, above) rear up and expand the hood at the back of the neck if they feel threatened or are excited.

THE TRUE COBRAS of the genus *Naja*, from the Sanskrit word naga, meaning snake, average 6–7 feet (1.8–2.1 m) in length. The Indian cobra has a dark body encircled by a series of light rings and, like all cobras, it has a characteristic hood behind the neck. A cobra flattens its hood horizontally by swinging out its long, moveable ribs to stretch the loose skin of the neck, in the way that the ribs of an umbrella stretch out the fabric. Cobras rear up and expand the hood if frightened or excited.

Wide-ranging species

Cobras are found in Africa and Asia, although the fossils of at least one species have been discovered in Europe, presumably dating from a time when the climate in that region was warmer. There are about 15 species of cobra, nine of which occur in Asia. However, the precise number of species varies according to the classification criteria used by different authorities.

The Indian cobra ranges from the Caspian Sea east across Asia, south of the Himalayas to southern China and the Philippines, and south to Bali in Indonesia. Throughout this range, the species' markings vary. In the west its hood has

typical "spectacle" markings, but toward the eastern side of India a single ringlike marking becomes more common, while in the Kashmir and Caspian regions the hood is marked with black transverse bars. The patterns are on the back of the hood but can be seen from the front because the stretched skin is translucent.

Among the African species are the Cape cobra, forest cobra, spitting cobra, Mozambique spitting cobra, and Egyptian cobra, which is also found in Asia. Some cobras, such as the Egyptian cobra, may be active by both day and night, the cycle of activity often depending on the season or the geographical location. Other species, including the Indian, spitting and forest cobras, are mainly nocturnal. The spitting and forest cobras are often associated with water.

Highly venomous

Cobra venom is secreted from glands that lie just behind the eyes. It runs down ducts to the fangs that grow from the front of the upper jaw. Each fang has a canal along its front edge, and in some species the sides of the canal fold over to form a hollow tube like a hypodermic needle, resembling the hollow fangs of vipers. A cobra strikes

COBRAS

CLASS	**Reptilia**
ORDER	**Squamata**
SUBORDER	**Serpentes**
FAMILY	**Elapidae**
GENUS	***Naja***

SPECIES   **15, including Indian cobra, *N. naja*; Egyptian cobra, *N. haje*; forest cobra, *N. melanoleuca*; Cape cobra, *N. nivea*; spitting cobra, *N. nigricollis*; and Mozambique spitting cobra, *N. mossambica***

LENGTH
Usually 6–7 ft. (1.8–2.1 m); some species (especially *N. melanoleuca*) may exceed 10 ft. (3 m)

DISTINCTIVE FEATURES
Stocky body; narrow head; unique hood behind neck that is expanded by flattening neck ribs; many species have spectacle pattern on hood

DIET
Mainly small mammals; also other snakes, frogs, toads, birds, large insects and (some species only) fish

BREEDING
***N. naja*: number of eggs: usually 12 to 20; hatching period: 50–60 days**

LIFE SPAN
Not known

HABITAT
Varied, from forest and open woodland to scrub and semiarid desert

DISTRIBUTION
Africa, except Sahara region, east through central Middle East to much of southern Asia and Southeast Asia

STATUS
Generally common

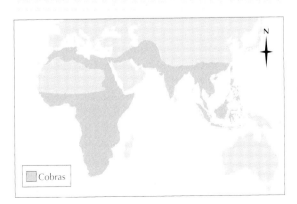

Cobras

upward, with the snout curled back so that the fangs protrude. As soon as the fangs pierce the victim's skin, venom is squirted down them by muscles that squeeze the venom gland.

Cobra venom acts on the nervous system causing paralysis, nausea, difficulty in breathing and, potentially, death through heart and breathing failure. The cobras' fangs are fairly short, but after a snake has struck it hangs on, chewing at the wound and injecting large quantities of venom. The seriousness of a bite depends on how long the cobra is allowed to chew.

The Indian cobra is regarded by many experts as being one of the most dangerous snakes in the world; its venom can kill within 15 minutes of a bite. In India figures of 10,000

All of the cobras are potentially dangerous to humans. The bite of the Cape cobra can cause death within 30 minutes.

Spitting cobras, such as this Mozambique spitting cobra, are the only snakes able to project their venom.

deaths a year from Indian cobra bites have been reported, which represents 1 in 30,000 of the population. Snakebite is common in Asia and Africa because many people go about barefooted in the countryside. Some cobras, notably the spitting cobra, defend themselves by spitting venom over a distance of up to 12 feet (3.5 m). They aim for the face of the attacker and the venom causes great pain and temporary or even permanent blindness if it enters the eyes.

Mating dances

Before mating, a pair of cobras "dance," raising their heads a foot or more off the ground and weaving to and fro. This may continue for an hour before mating takes place, when the male presses his cloaca (the chamber into which the reproductive, urinary and intestinal canals empty) against that of the female.

The Cape cobra mates between September and October and the eggs are laid a month later. These dates vary through the cobra's range as copulation and egg-laying takes place during the season most likely to provide abundant food for the young. Eggs number 8 to 20, and are laid in a hole in the ground or in a tree.

Female cobras may stand guard and are irritable and aggressive during the breeding period. A female is liable to attack without provocation, with potentially fatal consequences for passersby if her nest is near a footpath. Newly hatched cobras measure 10 inches (25 cm) in length.

Snake charmer's bluff

Cobras, especially the Indian and Egyptian species, are frequently used in snake-charming acts. It is generally understood now that the snakes are not reacting to the music but to the charmer's rhythmic movements. The pipe is merely a stage prop, and is not used by all performers because snakes are deaf, in the sense that they cannot perceive airborne vibrations. Most terrestrial animals have an eardrum that vibrates in time to airborne waves, and a system of bones and ducts that convey the vibrations from the eardrum to the sense cells of the inner ear. Cobras have neither. They can, however, detect vibrations through the ground. When the basket is opened, the snake is exposed to the glare of daylight. Half-blinded and shocked, it rears up in the defensive position with hood inflated, its attention caught by the swaying snake charmer.

COCKATOO

COCKATOOS HAVE CRESTS that they can erect at will, distinguishing the 21 species from the rest of the parrot family. Most cockatoos are white, sometimes with pink or yellow tinges and colored crests. Others, such as the palm cockatoo, *Probosciger aterrimus*, are all black. This species also has a bare patch of orange-pink skin on the cheeks that blushes red when the bird is excited. The male gang-gang cockatoo, *Callocephalon fimbriatum*, has a dark gray body with a striking scarlet crest, while the galah, *Cacatua roseicapilla*, or rose-breasted cockatoo, has a pale gray back and underwings and a rosy breast. The little corella, *Cacatua sanguinea*, was once known by the apt but descriptive name of "bloodstained cockatoo." The name cockatoo is derived from the birds' raucous calls.

Among people that live outside Australia by far the best known species of cockatoo are the sulfur-crested cockatoo, *Cacatua galerita*, or white cockatoo, and the cockatiel, *Nymphicus hollandicus*. These two species are common cage birds that breed freely in captivity.

Native to Australasia

Cockatoos are found in the Australasian region from the Celebes Archipelago in the west to the Solomon Islands in the east. They are primarily birds of forest and woodland though some are found in more open country. The cockatiel, for instance, is found in stands of trees bordering rivers and in the open scrub of the Australian interior. Outside the breeding season it is highly nomadic, living in flocks of up to several thousand that roam the country, settling where food and water are plentiful.

Although the cockatiel feeds in the open, it retires to woods or forests to roost at night, as does the palm cockatoo. The latter species sleeps alone, each bird having a roost on a high, usually bare branch. Unlike most other birds, palm cockatoos do not leave their roost until the sun is quite high. They then congregate in one tree and display, raising their crests and bowing to one another with wings extended and tails raised. As more palm cockatoos join the gathering the displays become more intense and the birds more excited. Each

bow is accompanied by a loud call of two notes, the second being shrill and drawn out. Eventually the noisy party of up to 30 birds sets out to feed, returning to shelter from the midday heat and finally retiring at nightfall.

While flocks of cockatoos are feeding, a sentinel is posted on a nearby tree, ready to give the alarm if a predator approaches. Having stood guard for a time, the sentinel flies down to the

Cockatoos are large, mainly white or mainly black parrots with huge bills and mobile crests. This is a pink cockatoo, Cacatua leadbeateri.

PALM COCKATOO

CLASS	**Aves**
ORDER	**Psittaciformes**
FAMILY	**Cacatuidae**
GENUS AND SPECIES	***Probosciger aterrimus***

ALTERNATIVE NAMES
Great palm cockatoo, great black cockatoo, goliath cockatoo, Cape York cockatoo

WEIGHT
1–2 lb. (0.5–1 kg)

LENGTH
Head to tail: 22–23½ in. (55–60 cm)

DISTINCTIVE FEATURES
Adult: black plumage; patch of orange-pink or scarlet, naked skin on each cheek; long, curved crest; massive, gray-black bill. Young: yellow tinge to underparts; pale bill.

DIET
Mainly seeds, nuts, berries and fruits, including thick-shelled palm fruits; some insect larvae

BREEDING
Age at first breeding: 1–5 years; breeding season: eggs laid July–March; number of eggs: usually 1; incubation period: about 33 days; fledging period: 78–81 days; breeding interval: 1 year

LIFE SPAN
Up to 50 years

HABITAT
Tropical rain forest and adjacent eucalypt woodland; wooded savanna

DISTRIBUTION
New Guinea, except mountainous areas; Aru Islands; Cape York Peninsula in northern Queensland, Australia

STATUS
Generally scarce; populations declining

Palm cockatoo

Several cockatoos are so abundant that they have become farmland pests. Others, among them the Moluccan cockatoo, Cacatua moluccensis, *are becoming scarce.*

flock to begin feeding and another cockatoo takes its place. This behavior is common in social bird species that feed on the ground. The predators of cockatoos include certain birds of prey and, more rarely, large lizards and snakes.

Gifted mimics

There has been extensive trapping of cockatoo populations because of the birds' popularity as pets. Except in the rare and localized species, however, such trapping is unlikely to have any effect on overall numbers. Apart from their colorful plumage, cockatoos are popular for their mimicking abilities and intelligence. Like other members of the parrot family cockatoos can learn to mimic a wide range of human words and animal noises, such as the clucking of chickens and the barking of dogs, as well as mechanical sounds. Cockatoos can also learn to perform tricks, including bowing and shaking hands.

There are frequent reports of captive cockatoos escaping to live in the wild in Europe and North America. However, the sudden change of environment, difficulty of finding appropriate food and threat from unfamiliar predators means that these birds are rarely able to survive outside captivity for long.

Nutcrackers

Cockatoos are mainly vegetarian, feeding on seeds, fruits and nuts. Cereal crops are a rich source of food to hungry flocks, which trample the plants and take the seeds, making them a major pest to farmers. The main pest species are the cockatiel, galah, little corella and sulfur-crested cockatoo. Other cockatoos, particularly the white-tailed black cockatoo, *Calyptorhynchus baudinii*, attack almonds, apples and pears, which are torn apart to reach the seeds.

Some species of cockatoos perform a useful service, by feeding on plants that cause paralysis and loss of sight in domestic animals. Moreover, during droughts the white-tailed black cockatoo and yellow-tailed black cockatoo, *Calyptorhynchus funereus*, help to make leaves accessible for hungry flocks of sheep. The cockatoos break off many twigs of scrubland plants, some of which fall to the ground, and the sheep follow the birds to strip the leaves from the fallen twigs.

Hole nesters

Some cockatoos, such as the little corella and the galah, breed in colonies; others, such as the yellow-tailed black cockatoo, nest alone. Cockatoos nest in holes in trees and cliffs, and corellas sometimes use hollows in termite mounds. The hole is lined with leaves or wood chips on which the eggs, usually four or five in number, are laid. Galahs sometimes strip the bark off the nesting tree to make a ring of smooth wood below the nest hole. This might be to prevent egg-eating lizards from climbing up to the nest.

The eggs are incubated for 3–4 weeks by both sexes and the nestlings fed in the nest hole until fledged. The cockatiel has been reported to wet its feathers in a nearby pool or river before entering the nest for a spell of incubation. Presumably this helps to keep the air in the nest moist and cool during the heat of the day.

Conservation

Seven species of cockatoos are now threatened, and several more, including the palm cockatoo, are considered to be near threatened. Most at risk are the critically endangered Philippine cockatoo, *Cacatua haematuropygia*, and the endangered yellow-crested cockatoo, *Cacatua sulphurea*, of Indonesia. The main causes of these declines are habitat loss and trapping for the pet trade.

The palm cockatoo nests in isolated pairs in tree holes about 30 feet (9 m) above the ground. Other species of cockatoos nest in loose groups.

COCKCHAFER

The cockchafer belongs to a very large family of beetles, the chafers. These insects have distinctive fanlike clubs at the tips of the antennae.

THE CHAFER BEETLES are plant-eating members of the scarab family, related to the dung beetles. Many chafers are considered pests as they cause untold damage to trees and other plants. Most species live in the Tropics.

The cockchafer, *Melolontha melolontha*, is a large European beetle that grows up to 1½ inches (3.5 cm) long. It has a square shape and a broad and deep abdomen. The elytra, or wing cases, are reddish brown and do not cover all of the abdomen, the tip of which protrudes like a conical tail. The head and thorax are black, and under the body there is a dense layer of hairlike bristles.

Emerge in early summer

The term cockchafer refers only to a single species, *M. melolontha*. The other members of *Melolontha* and those of several closely related genera are called simply chafers. Many of these beetles are popularly known as May beetles, May bugs, June beetles and June bugs, particularly in North America. These names derive from the fact that, although the adults emerge from pupation in about October, they do not start to fly until early summer. Throughout the summer months these insects are often observed striking lighted windows or car windshields.

Cockchafers can be seen in large numbers on fine evenings as the light is fading, sometimes climbing up grass stalks and pumping their abdomens in and out, presumably to war body before takeoff. They then fly off th the grass with a slow, meandering flight. flying cockchafers produce an audible hun noise. However, the wingbeat of a cockch. relatively slow, at 45 beats per second, compared with 130–240 beats per second typical species of bumblebee.

Sensory antennae

One characteristic feature of chafers and relatives is their elaborate antennae, whic in a fan of thin plates. These structures given the superfamily to which chafers the name Lamellicornia, or leaf-horns, be the antennae resemble the leaves of a opened book. Antennae are sensory on reacting to sound and smell. Examination insect's antennae is an easy way for scient determine the acuteness of that species' he and smell. If the antennae are well-deve then the chances are that the insect's hearin smell are sensitive. Chafers probably use fanned antennae to find food.

Pests throughout life cycle

Both larval and adult chafers damage plant larvae attack the roots while the adults e leaves and petals or suck sap and nectar. chafers feed on tree foliage, and when abu can cause extensive damage by stripping all the leaves. Among the most serious pests are the leaf chafers: any one of more than 4,000 species in the subfamily Rutelinae. The rose chafer, *Cetonia aurata*, a common day-flying species, favors the foliage of grape plants and the petals of roses. The larvae of the masked chafers, *Cyclocephala boealis* and *C. lurida*, attack turf and root crops in much of the United States, causing thousands of dollars of damage each year.

Burrowing larvae

Female chafers lay several batches of eggs in the early summer. In the cockchafer each batch numbers 12 to 30 eggs, totaling about 70 in all. The eggs are laid in burrows 6–8 inches (15–20 cm) deep, and hatch within 3 weeks. The larvae of some chafers live in rotten trees and logs through which they burrow, grinding up the wood with their strong, horny jaws. Wood is digested with the help of gut bacteria that can break down cellulose, its main constituent. Burrowing chafer larvae are probably important in assisting the breakdown and conversion of dead trees into humus, but they also cause considerable damage to commercial timber plantations.

Cockchafer larvae spend 3–4 years in the soil before pupating, the exact length of the period depending on the climate. Each winter the larvae burrow down to avoid frosts, returning to the surface in spring to feed on the roots of grasses and other plants. At the end of their third summer the larvae burrow down and pupate in a cocoon some 2–3 feet (60–90 cm) beneath the surface. They emerge as adult cockchafers a few months later. The larvae are sometimes called rook-worms because the rook, *Corvus frugilegus*, a Eurasian crow, is one of their main predators.

Chafers sometimes occur in vast numbers, and the swarms can have a dramatic impact on an area's vegetation and timber.

CHAFERS

PHYLUM	**Arthropoda**
CLASS	**Insecta**
ORDER	**Coleoptera**
FAMILY	**Scarabaeidae**
GENUS	***Melolontha, Cetonia, Phyllopertha, Eupoecilia, Cotinis, Cyclocephala*, others**
SPECIES	**Cockchafer, *M. melolontha*; rose chafer, *Cetonia aurata*; green June bug, *Cotinis mutabilis*; garden chafer, *P. horticola*; many others**

ALTERNATIVE NAMES
May beetle; May bug; June beetle; June bug; common cockchafer (*M. melolontha* only); green fig-eater (*C. mutabilis* only)

LENGTH
Up to 1½ in. (3.5 cm)

DISTINCTIVE FEATURES
Large, square-shaped body; elaborate, fanned antennae; elytra (wing cases) do not cover the entire abdomen; thus the tip is exposed

DIET
Adult: leaves, buds, flowers and sap. Larva: roots, especially of cereals and other grasses; wood.

BREEDING
Varies according to species; hatching period: about 20 days

LIFE SPAN
Larva: up to 3–4 years

HABITAT
Most well-vegetated habitats, including forest, farmland, grassland and gardens

DISTRIBUTION
Almost worldwide

STATUS
Often abundant

COCKLE

Empty cockle shells are often washed up onto beaches by high tides and storms.

Abundant in sheltered bays

There are approximately 250 species of cockle in the superfamily Cardiacea, and they have a worldwide distribution. The edible cockle is found from midtide level down to a depth of 8,000 feet (2,450 m), lying obliquely not more than 2 inches (5 cm) below the surface of clean sand, mud or muddy gravel. Generally the average size increases from high to low water. Cockles are particularly common in the sheltered waters of bays and the mouths of estuaries, as many as 10,000 having been recorded per square meter. Although cockles do live upstream in estuaries, the largest are found far from fresh water. Dilution of the salt water with fresh water has the effect of producing a less regularly shaped, thinner shell, with fewer ribs.

Structure of cockles

It has been suggested that the 20 to 24 ribs which radiate out from a cockle's hinge region help to hold it firmly in the mud or sand, and that the globular shape helps to protect the cockle should it be dislodged and rolled about by wave action. A ribbed and more rounded shell is much stronger than a smooth or flat shell. This may be an adaptation to avoid being eaten by predators that crush shells, such as some fish.

At the back of the body is a pair of short tubes, or siphons, joined at the base. Water containing oxygen and the plankton and organic matter which are the food of the cockle, is drawn in through the lower siphon, the one farthest from the hinge. An outgoing jet sweeps through the upper siphon, carrying away wastes.

The siphons are the only parts of the buried mollusk to project above the seabed, and so they are an appropriate site for the eyes. These are small but complex, with retinas and lenses, and are mounted on sensory tentacles. The eyes enable the tentacles to be withdrawn if a shadow falls on them, so protecting them from predators, but presumably eyes can have little other use in such a sedentary, filter-feeding animal.

Each cockle can use its long, orange foot to flip itself up in the water. The jumping cockle, *L. crassum*, is the most notable leaper.

Conveyor-belt digestion

Cockles feed on small plant particles that abound in the water: single-celled plants such as diatoms and the spores and fragments of larger algae. There is little selection of what is eaten, however, and a cockle's stomach usually contains much sand and mud as well. The water bearing these

THE COCKLE IS AN unusually globular bivalve mollusk; the two valves, or shells, are ribbed and similar in shape. Cockles range in size from just over ⅓ inch (1 cm) in diameter to about 6 inches (15 cm), the size of the smooth giant cockle, *Laevicardium elatum*, of California, and the giant Pacific cockle, *Trachycardium quadragenarium*, among others.

The prickly cockle, *T. egmontianum*, and the yellow cockle, *T. muricatum*, are both plentiful on the Atlantic coast of North America, while the common Pacific egg cockle, *L. substriatum*, and the fucan cockle, *Clinocardium fucanum*, both under 1 inch (2.5 cm) in diameter, occur along North America's Pacific coast.

Of the various cockles native to the coasts of western Europe, by far the most common is the edible cockle. Although its present-day scientific name is *Cerastoderma edule*, it is often referred to by the older name of *Cardium edule*. The latter name reflects the heart-shaped appearance of the pair of shells when viewed end-on, and also the edibility of their contents.

COMMON COCKLE

PHYLUM	**Mollusca**
CLASS	**Bivalvia**
SUBCLASS	**Lamellibranchia**
ORDER	**Heterodonta**
FAMILY	**Cardiidae**
GENUS AND SPECIES	***Cerastoderma edule***

ALTERNATIVE NAME
Edible cockle

LENGTH
Up to 2½ in. (6 cm)

DISTINCTIVE FEATURES
Pair of well-rounded, heart-shaped shells, each with up to 24 broad ribs; annual growth rings run across ribs and provide approximate guide to age

DIET
Filters small planktonic organisms and particles from water

BREEDING
Breeding season: spring and summer; number of eggs: variable; breeding interval: 1 year

LIFE SPAN
Up to 6–7 years

HABITAT
Mud, sand and fine gravel in coastal waters, from midtide mark to depth of 8,000 ft. (2,450 m); most abundant in sheltered bays and estuaries

DISTRIBUTION
Coasts of eastern Atlantic, from northern Norway south to West Africa; also present, but less numerous, on Mediterranean coasts

STATUS
Generally abundant, sometimes superabundant

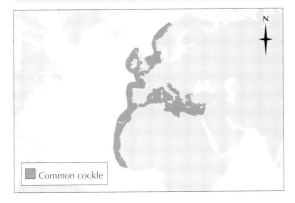

Common cockle

particles enters through the lower siphon and is expelled by the upper siphon after passing through a fine latticework of sievelike gills. The cockle's cilia, as well as creating this current, also propel the trapped food toward the mouth, itself flanked by a pair of cilia-covered lips.

In the stomach is a curious structure found only in mollusks: a rotating coiled rod up to 1 inch (2.5 cm) long called a crystalline style, turned by cilia lining the pocket which secretes it. The crystalline style, made up of digestive enzymes, gradually wears away and dissolves at the tip. Its rotation serves to draw along the food particles trapped in strings of mucus and at the same time its enzymes help to digest the food.

Breeding left to chance

Though some related species are hermaphrodites, the sexes are separate in the edible cockle. Spawning begins at the end of February or early in March and continues until June or July, the eggs and milt being released freely into the water. The eggs that become fertilized by this uncertain method develop into minute, free-swimming larvae, called spats, propelled by the beating of cilia. Eventually each larva develops a shell and foot and the resulting young cockles, still less than 1 millimeter long, settle on the sea bottom. The success of spawning varies from year to year and populations tend to be maintained by particularly good "spat falls" perhaps once in 3 or 4 years.

As the shells grow, fine concentric grooves appear, marking former positions of the shell edge and providing a guide to age. A full-grown cockle has three to seven grooves. The groove reflects the slower growth of the shell in winter when lack of sunshine reduces plankton levels.

The cockle is a bivalve mollusk, its two shells, or valves, being hinged toward its rear. The shells are held together by two muscles known as adductors, visible on either side of the hinge.

COCK-OF-THE-ROCK

Cock-of-the-rock have helmet-shaped crests unique in the bird world. This is a male of the Andean species.

THE COCK-OF-THE-ROCK are two ornate and rather uncommon South American species in the cotinga family, which is noted for the variety of wattles and crests worn by its members. The pair of cock-of-the-rock are immediately recognizable by bushy crests that extend as a complete double fan from the base of the bill to the crown of the head. They are jay-sized birds, about 1 foot (30 cm) long, with short, inconspicuous tails. The Guianan cock-of-the-rock is orange over most of its body except for the black and pale gray flight feathers on the wings and a narrow chestnut stripe running around the margin of the crest. The scarcer Andean cock-of-the-rock is a lighter, reddish orange with black on the wings and tail. The females are drab brown, the female Guianan cock-of-the-rock being darker.

Confined to humid forests

Cock-of-the-rock are found in dense, humid forests in and around the Amazon basin. The Guianan cock-of-the-rock occurs in Suriname, French Guiana, Guyana, southeastern Venezuela and northern Brazil. The other species is found farther to the west and south, on either side of the Andes mountain chain. The most likely place to find cock-of-the-rock is in damp forest with plenty of rocky outcrops, which provide shallow caverns for nesting. The birds live near the floor of the forest rather than in the canopy, and have the strong legs and claws associated with species that spend much time on or near the ground.

Cotingas are mainly fruit eaters, and cock-of-the-rock are no exception, but they also feed on insects, spending a large part of their time searching for prey on the ground. When cock-of-the-rock have been kept in captivity it has been found that they quickly lose condition and die if insect food has not been provided. Besides insects, captive birds have been found to relish snails, which they open by smashing the shells against rocks.

Spectacular displays

The cock-of-the-rock are among the many birds in which the male takes the minimum part in breeding. Male cock-of-the-rock have nothing at all to do with nesting, once they have mated. They instead devote significant time and energy to performing highly elaborate display rituals that are designed to show off their ornate plumage. During these communal displays the males challenge rivals and beckon females.

A displaying male cock-of-the-rock appears so unusual to human eyes that at first sight it is hard to believe that it is a bird. The bill and tail become covered by the crest and other ornate plumage, obscuring the form of the body. When the display starts, the crest is fanned out so that the two halves spread over the bill. At the same time, the feathers of the back and breast are fluffed out and the tail coverts (the feathers at the base of the tail) are fanned out. These feathers are fluffy and stand out, so the bird looks as if it is wrapped in a shawl.

Male cock-of-the-rock have traditional displaying grounds, or arenas, where as many as 40 birds regularly gather in the breeding season. Within the arena each bird has a private perch on a branch near the ground and a patch of cleared ground about 1 foot (30 cm) across. The ground in the display arena is cleared of dead leaves by the draughts caused by the owners landing and taking off. As well as displaying their brilliant plumage, the male cock-of-the-rock have a variety of calls and movements. They bob their heads, showing off the crest, and hop about the perches, flicking their wings. Together with buglelike calls uttered as the birds arrive at the arena, the males make two mechanical noises. One is a finger-snapping sound made with the bill as the head bobs and the other is made when the birds are chasing one another on the wing.

COCK-OF-THE-ROCK

CLASS	**Aves**
ORDER	**Passeriformes**
FAMILY	**Cotingidae**
GENUS AND SPECIES	**Andean cock-of-the-rock,** *Rupicola peruviana*; **Guianan cock-of-the-rock,** *R. rupicola*

LENGTH
Head to tail: *R. peruviana*, 12 in. (30 cm); *R. rupicola*, 10½ in. (27 cm)

DISTINCTIVE FEATURES
Stocky body; short bill and tail; laterally compressed, fan-shaped crest, smaller in female. *R. peruviana*: **(male) bright reddish orange with black wings and tail; (female) duller, brownish orange.** *R. rupicola*: **(male) mainly intense orange with black and gray wings; (female) dark, dusky brown.**

DIET
Mainly fruits; some invertebrates

BREEDING
Breeding season: March–June; number of eggs: 2; incubation period: probably 25–28 days; fledging period: probably 40–44 days

LIFE SPAN
Not known

HABITAT
R. peruviana: **forested ravines and streams in mountains; up to 8,000 ft. (2,400 m).** *R. rupicola*: **lowland tropical forest near rocky outcrops; below 4,000 ft. (1,200 m).**

DISTRIBUTION
R. peruviana: **Andes Mountains, from western Venezuela south to western Bolivia.** *R. rupicola*: **eastern Venezuela east to Suriname, Guyana and French Guiana and south to northern Brazil.**

STATUS
Uncommon and highly localized within respective ranges

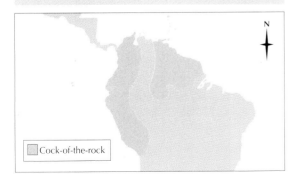

Cock-of-the-rock

One of the wing feathers has a modified tip that vibrates to produce a whistling sound when the bird is in flight.

When a female cock-of-the-rock approaches the arena the males become excited and soon begin to display, together producing a veritable cacophany of bright color, rapid motion and weird sounds. The female watches the frenzied displays for a few minutes then flies down to the male of her choice, hitting the ground by him and immediately taking off again. He follows and they mate a short distance away.

Females build their nests in caves and under overhangs on cliffs. Several nests may be built within a few feet of each other and the females roost on the nests even when not incubating or brooding. The nests are made of mud bound

with saliva, vines and roots. Two eggs are laid and the fledging period seems unusually long but, because of the species' inaccessible habitat, little is known for certain about parental care.

Neighbors drift apart
The Guianan cock-of-the-rock is presumed to be the ancestor of the Andean species. Since the latter became separated from its close relative it has diverged into two subspecies that live on each side of the Andes Mountains. *Rupicola peruviana sanguinolenta* occurs to the west, while *R. p. aurea* lives in the east. The male sanguinolenta has developed fiery, blood-red plumage and its iris is mainly red, rather than yellowish or bluish white. The differences between the subspecies of *R. peruviana* have been made possible by the complete isolation of the two populations by the cold climate of the Andes range. If there had been passes through the mountains, low enough to allow a continuous belt of tropical vegetation to link east and west, the populations would have mixed and interbred.

Male cock-of-the-rock (R. peruviana, above) compete for females at arenas. Strong males with impressive dancing skills will win several partners, weaker males often do not mate at all.

COCKROACH

Cockroaches (adult American cockroach, above) eat all manner of human foodstuffs as well as a wide range of substances not usually regarded as edible, such as glue, electrical cable insulation and soap.

COCKROACHES ARE FAIRLY large insects, flat in shape and with two pairs of wings, the forewings being more or less thickened and leatherlike. These serve as a protective cover for the delicate hind wings, just as the hardened forewings, or elytra, of beetles do. The hind wings of cockroaches are pleated like a fan when not in use; when expanded for flight they have a very large surface area. Cockroaches have triangular-shaped heads and their mouthparts are angled downward. They also possess a pair of long and highly mobile antennae. These are sensitive to touch and enable one cockroach to smell and taste subtle chemical emanations from another, making sexual recognition possible.

Classification

Fossil discoveries have revealed that there were already many species and abundant populations of cockroaches at the time that coal deposits were formed, 300 million years ago, in the Carboniferous period. The Carboniferous cockroaches appear to have differed little from present-day species.

Cockroaches were at one time classified with grasshoppers, crickets and stick insects in one large order, the Orthoptera. This order has now been split up into several separate orders. One of these, the Dictyoptera, is comprised of the praying mantises and the cockroaches.

Colonizers of human environments

The most familiar of the nearly 4,000 species of cockroach are the tropical and subtropical forms that have taken advantage of the warmth and the opportunities for scavenging afforded by homes and premises in which food is made or stored. By this means cockroaches have extended their range into temperate and cold regions, and some species have been artificially distributed all over the world by human commercial activities. In the natural state, the great majority of cockroach species are confined to the Tropics.

Among the most abundant species of cockroach is the common cockroach, *Blatta orientalis*, or black beetle. It is of variable length, averaging about 1 inch (2.5 cm) and is dark brown; the adult females are almost black. Both sexes are

492

COCKROACHES

PHYLUM	**Arthropoda**
CLASS	**Insecta**
ORDER	**Dictyoptera**
FAMILY	**Blattellidae, Blattidae**
GENUS	**Many, including *Periplaneta*, *Blatta*, *Blattella*, *Gromphadorhina* and *Ectobius***
SPECIES	**Almost 4,000, including American cockroach, *P. americana*; German cockroach, *Blattella germanica*; common cockroach, *Blatta orientalis*; and giant Madagascar hissing cockroach, *G. portentosa***

ALTERNATIVE NAMES

Black beetle (*B. orientalis* only); steamfly, shiner (*B. germanica* only)

LENGTH

Varies according to species; *B. germanica*: ½ in. (13 mm); *P. americana*: 1¼–2½ in. (3–6 cm); *G. portentosa*: up to 4 in. (10 cm)

DISTINCTIVE FEATURES

Flattened body with enlarged thorax; head largely hidden when viewed from above; mouthparts face downward; long antennae; long legs covered with protruding hairs; winged species have thick forewings that cover membranous hind wings; most species brown in color

DIET

Almost anything digestible; dead animal material probably the main food in wild

BREEDING

Age at first breeding: 5–15 months; breeding season: all year; number of eggs: 12 to 50; hatching period: few weeks to several months, according to species and environment; breeding interval: 3 or 4 generations per year in *B. germanica*

LIFE SPAN

Usually 1 year; some species up to 2–4 years

HABITAT

Wide variety of terrestrial habitats and human environments

DISTRIBUTION

Most species in Tropics; *P. americana*, *Blatta orientalis* and *Blattella germanica* have virtually worldwide distribution as result of human commercial activities

STATUS

Many species common or abundant

flightless: the wings do not reach the tip of the body in the male and are vestigial in the female. The common cockroach is a familiar domestic pest, though its region of origin is unknown.

The German cockroach, *Blatella germanica*, also known as the steamfly or shiner, is about ½ inch (13 mm) long and yellowish brown in color with two dark brown stripes on the prothorax (forepart of the thorax). Its wings are fully developed. This species is almost as abundant as the common cockroach. Despite its name, it is certainly not of German origin and is probably a native of North Africa.

The American cockroach, *Periplaneta americana*, is reddish brown overall and also has fully developed wings. It is often found in seaport towns and on ships, but in tropical countries it is the chief house-dwelling cockroach. The species, like the German cockroach, probably originated in North Africa.

In common with termites, some species of cockroach are able to digest wood, with the aid of microscopic animals that live in their intestines.

Omnivorous scavengers

In the wild, most cockroaches scavenge dead animal remains, fallen fruit and fungi. Some species feed on wood, which they digest with the help of protozoans, microscopic one-celled animals, found in their intestines. Termites, which are closely related to cockroaches, eat and digest wood by the same means.

In human habitations, cockroaches eat any kind of human foodstuffs available. As true omnivores, they will also eat substances that are not generally regarded as edible, including book-bindings, the plastic coverings of electric cables, boot-blacking, soap, ink and whitewash. Cockroaches also feed on other cockroaches. The harm cockroaches do around the house is greatly increased by their habit of fouling, with their droppings, far more than they actually eat. This habit of excreting while feeding has seen cockroaches become widely regarded as a pest by humans. However, cockroaches are not known to convey any disease and only about 1 percent of all species live in close contact with humans.

Breeding

Cockroach eggs are enclosed in a purselike capsule called an ootheca. In the common cockroach this is carried for a day or two, protruding from the body of the female, and then dropped, or sometimes stuck in a crevice. Some species of cockroaches carry their eggs internally for the full hatching period. The female cockroach often cares for the ootheca until the eggs hatch, even if she does not carry it with her. The egg sac is white when it first appears at the tip of the abdomen, but darkens later and, when deposited, is almost black and less than ½ inch (13 mm) long.

Normally an ootheca contains 16 eggs in two rows of eight, though the exact numbers vary. The eggs hatch 2–3 months after the formation of the ootheca, which splits to allow the young cockroaches, or nymphs, to emerge. These are about 5 millimeters long on hatching, and white, gradually becoming brown as they grow. The nymphs resemble the adults in form, except that the wings are absent, and take up to 15 months to reach maturity. Molting of the exoskeleton, or ecdysis, takes place 6 to 12 times in the course of the metamorphosis from nymph to adult. The breeding habits of the American cockroach are very similar to those of the common cockroach.

In the German cockroach the ootheca is carried by the female for up to 24 hours before hatching, although the eggs may hatch while it is still attached to her. The egg sac is chestnut brown in color and, a few days before hatching, a green band appears along each side of it. Hatching usually takes place 4–6 weeks after the ootheca is formed and the egg sac contains 35 to 45 eggs.

Female cockroaches encourage their newly hatched young to follow them in search of food, even though the nymphs' limbs are not yet very strong. If the nymphs sense danger, they immediately seek refuge behind their mother's long legs, or under her wings.

Cockroach nymphs (common cockroach, below) resemble adults in form, but lack wings. During the course of their development into adults they may shed their exoskeleton up to 12 times.

COD

THE COD IS THE second most valuable food fish in the world, after the herring. The cod family contains 150 species, of which about 10 are important food fish, including the coalfish or pollack, haddock, hake and whiting. Cod are marine fish, though there is one entirely freshwater species, the burbot, *Lota lota*, found in northern parts of Eurasia and North America.

The Atlantic cod is round-bodied, up to 100 pounds (45 kg) in weight and 6 feet (1.8 m) long, although those usually sold are 2–25 pounds (1–11 kg). Its typical color is olive green to brown, the back and flanks marbled with spots, the belly white. There are three dorsal and two anal fins, the snout projects over the mouth and there is a prominent, whiskerlike barbel on the chin.

Seasonal movements

The distribution of the Atlantic cod is reflected in the locations of the main fisheries, in the North Sea and off Norway, Bear Island, Iceland, Greenland and Newfoundland. The Arctic cod, *Boreogadus saida*, which is also known as the polar cod, is a slender-bodied species that lives in the northern Pacific.

Within each area the populations of Atlantic cod are self-contained, and several subspecies have been recognized. However, the larger, older fish make long migrations of several hundred miles between the areas, for example from Newfoundland to Greenland and from Greenland to Iceland. In each region the species has distinct seasonal feeding grounds and spawning grounds. In late spring and summer Atlantic cod live in deep waters, at depths of up to 650 feet (200 m). In the fall, the shoals move to shallower waters along coastlines; at times cod can be caught just 50 yards (45 m) offshore.

Atlantic cod also execute a daily movement related to the intensity of light, similar to that found in a number of other shoaling fish. Even when at depths of 600 feet (180 m), the cod form compact shoals during the daylight hours, disperse at sunset and reform at sunrise.

Voracious predators

Cod feed primarily on other fish, especially herring, mackerel and haddock, as well as sand eels and smaller members of their own species. Squid and bottom-living invertebrates such as shrimps, crabs, mollusks and worms are also eaten. Cod have strong, sharply-pointed teeth, and their digestive juices can dissolve seashells and the shells of crabs. The overlapping upper lip is an adaptation for scooping up worms and other prey from the seabed.

Vast numbers of eggs

There is no marked outward difference between the sexes of Atlantic cod, which become sexually mature at 4–5 years, when 2–3 feet (60–90 cm) long. In the first 3 months of the year adults of breeding age move to the spawning grounds. This may take the fish across very deep water, for example from Norway to Bear Island. The females shed their eggs and the males release milt into the sea; fertilization is random.

A large female cod lays up to 9 million eggs, each 1–2 millimeters in diameter, which float to the surface. Generally these hatch in about 10–20 days and the larvae remain in the surface plankton for the next 60–75 days. When just over ¾ inch (2 cm) long they move down to the bottom, into depths of about 250 feet (70 m), to feed on small crustaceans, amphipods, isopods and small crabs. By the end of the year the young cod are 6 inches (15 cm) long. Young cod of a similar age keep together, and are known as codling.

The Atlantic cod hunts fish and crustaceans in large groups. However, years of overfishing have severely reduced its numbers and the species is now vulnerable.

Cod feed mainly near the seabed, where the chin barbel is used to scoop up invertebrates.

Place in human history

Atlantic cod have been fished by humans since the 16th century and during the 20th century the annual catch reached 300 million to 400 million fish. As a source of revenue cod have had an important bearing on the course of human history. In the early 16th century, Spanish, Basque, French and English fishers caught cod in the North Sea and North Atlantic. By the middle of that century 300 French ships were fishing on the Grand Banks of Newfoundland. Many of the sailors crewing the fighting ships of the Spanish Armada learned seamanship in cod ships.

England was slow to exploit these "silver mines," as the seemingly inexhaustible supplies of cod off Newfoundland were then known. Early in the 17th century, however, there came a two-pronged attack. Ships from western and southern England were making the long journey and returning with valuable cargoes of salted cod. By 1634 an estimated 18,700 English seamen were working the Newfoundland fishery. The colonists of New England had also discovered this source of wealth and by 1635, one generation after the colony was founded, 24 vessels were exporting up to 300,000 cod a year.

Atlantic cod led the English and the French to Canada, and when the United States was founded a title to the cod-fishing grounds was included in the Act of Independence. The species appears on bank notes, seals, coins and revenue stamps of the New England colonies, and a carved figure of a cod still occupies a place of honor in the Massachusetts State House. The fish left its mark in the form of open strife between the crews of different nations.

Salted cod soon became a standby for seafarers, explorers and armies as Europeans settled the new-found continents. Nothing of the fish was wasted. Its skin yielded glue, its liver was used to produce a high-grade oil and its swim bladder furnished a semitransparent gelatin used as a clarifying agent and in glue and jellies.

COD

CLASS	**Osteichthyes**
ORDER	**Gadiformes**
FAMILY	**Gadidae**
GENUS AND SPECIES	***Gadus morhua***

ALTERNATIVE NAMES
Atlantic cod; codling (young only); codfish (archaic)

WEIGHT
Up to 100 lb. (45 kg), usually 5–25 lb. (2–11 kg)

LENGTH
Up to 6 ft. (1.8 m), usually under 3 ft. (0.9 m)

DISTINCTIVE FEATURES
Rounded body; prominent barbel on chin; 3 dorsal and 2 anal fins; body mainly greenish brown with dark mottling on sides, white lateral line and white underside

DIET
Adult: other fish including smaller cod; wide range of invertebrates.
Young: plankton, larvae and eggs.

BREEDING
Age at first breeding: usually 4–5 years; breeding season: February–April; number of eggs: up to 9 million; hatching period: 8–23 days at 37–52° F (3–14° C)

LIFE SPAN
Up to 20 years

HABITAT
Mainly near seabed; summer: usually in waters 500–650 ft. (150–200 m) deep; winter: in shallower waters nearer coast

DISTRIBUTION
North Atlantic, Baltic Sea and White Sea (Arctic waters to north of European Russia)

STATUS
Vulnerable; large declines in many areas

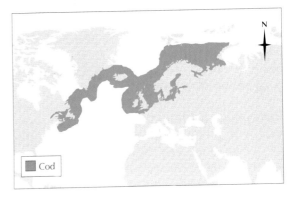

Cod

COELACANTH

IRST MADE KNOWN to western science in 1938, and belonging to an order previously thought to have become extinct 70 million years ago, the coelacanth is a 5-foot (1.5 m) long, 120-pound (55 kg) species of primitive fish. The word coelacanth (pronounced seelakanth) means "hollow spines," and refers to the spines of the fins. It now seems that fishers of the Madagascar region may have known of the coelacanth for thousands of years, by the name of kombessa. They regard the species as a poor food fish unless salted and dried, as its skin is slimy and when caught it continually oozes oil.

One of a kind
The coelacanth is of robust build, and brown to metallic dark blue in color. Its fins, tail, head and backbone and some of its internal organs are of unique structure.

The pectoral, pelvic and anal fins, instead of being set directly on the body, are carried on strong, muscular lobes that seem to be halfway between normal fins and the walking limbs of primitive land animals. There are two dorsal fins, the rear one also being lobed. In most fish the junction between the body and tail fin is marked by a constriction, but in the coelacanth the body narrows rapidly and evenly, and then continues backwards as a narrow strip. The rays of the tail fin are divided into two equal parts, one above and one below this narrow extension of the body; this double tail serves as a paddle.

Another peculiarity of the coelacanth is its hard scales. Each scale is a bony plate covered with dermal denticles (small, toothlike points) like those of sharks. This arrangement effectively provides armor plating.

Discovery of a living fossil
On December 22, 1938, a fishing boat dragging the seabed off the mouth of the Chalumna River, west of East London in South Africa, was unloading its catch when the crew came upon a fish they had not seen before. It lived for just 4 hours. The skipper, Captain Hendrick Goosen, sent the fish to Marjorie Courtenay-Latimer, curator at the East London museum, who immediately forwarded a description and sketches to

J. L. B. Smith, South Africa's foremost professor of ichthyology (the study of fish). By the time Smith finally saw the fish it had been gutted, stuffed and mounted. The remarkable discovery made headlines around the world, and the new species was named *Latimeria chalumnae*, after Courtenay-Latimer. The second coelacanth to be found was not caught until 14 years later, by a fisherman near the Comoros Islands, west of Madagascar. Since then more than 100 other specimens have been caught.

Coelacanths are known from the fossil record dating back over 360 million years, and seem to have been most common about 240 million years ago. They live in deep water, where their dark coloration provides camouflage, and appear to be associated with lava caves.

Scientific enigma
In addition to its strange external appearance the coelacanth has several peculiar internal features. Although it is a bony fish, the backbone is made up almost entirely of a large, tough cartilaginous rod, the notochord. In an evolutionary sense, the

The coelacanth is the last survivor of an ancient lineage of fish. It sculls forward with its limblike fins, which move in a manner similar to the legs of quadrupedal mammals.

497

The first coelacanth to be examined by scientists is preserved in the East London museum, South Africa. Its discovery in 1938 was one of the great zoological finds of the 20th century.

COELACANTH

CLASS	**Osteichthyes**
SUBCLASS	**Sarcopterygii**
ORDER	**Coelacanthiformes (or Actinistia)**
FAMILY	**Coelacanthidae**
GENUS AND SPECIES	***Latimeria chalumnae***

ALTERNATIVE NAME
Kombessa

WEIGHT
Up to 120 lb. (55 kg)

LENGTH
Up to 6 ft. (1.8 m)

DISTINCTIVE FEATURES
Robust body tapering at rear to form split, paddle-like tail; limblike pectoral, pelvic and anal fins set on muscular lobes; soft, hollow fin spines; thick, bony-plated scales; slimy to the touch; overall color blue mauve with metallic sheen and faint white spots

DIET
Probably other fish

BREEDING
Bears live young; gestation period: probably about 360–400 days; number of young: probably 5 to 25

LIFE SPAN
Not known

HABITAT
Tropical seas at depth of 500–2,300 ft. (150–700 m), especially over rocky seabeds with caves and in areas of strong current

DISTRIBUTION
Western Indian Ocean, mainly near coasts of Madagascar and Comoros Islands but sometimes off Mozambique and South Africa. Second population discovered off Sulawesi, Indonesia, in 1998 may involve new species.

STATUS
Endangered; little population data available

notochord evolved before the backbone; in the developing embryo of vertebrates it appears first and is later enclosed by the vertebrae and lost, except in primitive species, such as lampreys. The coelacanth's heart is simple, and resembles the type of organ predicted by anatomists when trying to explain how the heart evolved.

The coelacanth's kidneys, rather than lying just under the backbone, are located on the floor of the abdomen. Moreover, instead of being a pair they are joined. The stomach is also peculiar, being essentially a large bag. The intestine has a spiral valve, a feature shared with sharks and other primitive fish.

Further finds

Skin divers have searched for coelacanths off the coasts of Madagascar and neighboring islands, and some have caught a glimpse of large fish believed to be coelacanths, along the steep slopes where the rocky seabed suddenly dips in a vertical wall into very deep water. However, when brought to the surface, coelacanths soon die, due to the combination of decompression and exposure to warmer water.

The conservationist Hans Fricke filmed living coelacanths from a submersible at a depth of 560–660 feet (170–200 m). His observations revealed that the species occasionally stands on its head, possibly to stimulate an electro-detection mechanism in its snout. Contrary to early speculation, the fish does not use its fins to walk on the seabed.

A trade in the coelacanth sprang up after the 1950s, threatening the species' survival, but has since collapsed. Moreover, several coelacanths were found in Indonesia in the late 1990s, so the species may be more widespread than previously thought. Some authorities think that these fish may belong to a second species of coelacanth.

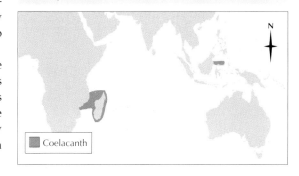

Coelacanth

COLLARED DOVE

ONE OF THE SMALLER DOVES, the collared dove is native to Eurasia but today is spreading throughout the southeastern United States. The species was introduced to Florida, near Miami, and on occasion has also bred as far east as Houston, Texas. It is a little larger than the turtle dove, *Streptopelia turtur*, which breeds in Eurasia and North Africa and winters in sub-Saharan Africa. The collared dove is most easily distinguished from its close relative by a narrow black collar around the back of its neck. It is otherwise pale gray-buff, with blue-gray shoulders that show in flight. Sometimes the collared dove is known as the collared turtle dove.

Huge range expansion

The collared dove is a bird of farmland, parks, gardens and villages, and in many regions is a common sight in city centers: Originally it was a bird of Asia and Africa, ranging from Korea and northern China, across India to the Middle East and the southern borders of the Sahara Desert.

Collared doves became established in southeastern Europe in the 18th century, but whether they spread there naturally or were introduced by Turkish people is not known. The species began to spread west-ward in the early 20th century, in increasing numbers from about 1930. By 1967 collared doves had reached the westernmost parts of Europe, and were breeding in Fair Isle, to the north of mainland Scotland, and in Ireland.

The dramatic range expansion of the collared dove was a steady movement to the northwest. In 1912 the species had reached in Belgrade, in the former Yugoslavia; by 1930 it was found in Hungary, and by 1943 it was breeding in Vienna, Austria. The collared dove reached Denmark in 1948 and Sweden a year later. In Britain the first collared doves to arrive bred in 1955. Since then the population, aided by immigrants, has rapidly expanded. By 1964 it was estimated at 19,000 birds. In the early 1990s the British population had grown to about 200,000 pairs, but numbers are now thought to have leveled off.

The collared dove is not the only species of bird to have extended its range across Europe in recent times. For example, two species of finch, the serin (*Serinus serinus*) and the scarlet rose-finch (*Carpodacus erythrinus*), have spread consid-erably, and the gray wagtail (*Motacilla cinerea*), a small pipit-like bird, spread into northern

Germany and Scandinavia during the 19th century. The basis for these range expansions is probably the general warming of the European climate. Winters have become slightly milder, so that nonmigratory birds like the collared dove can survive farther north during the winter months. The collared dove has also been helped by its ability to take advantage of human activity. It feeds on the ample supplies of spilt grain around farms and villages, and the rate of its spread may have been accelerated by the expansion of European agriculture.

Gleaning doves

For many collared doves, spilt grain lying on the ground represents a larger part of their diet than naturally occurring grass seeds. In winter the doves gather in flocks of a hundred or more to feed in stubble fields. The tips of grass shoots and seed heads are also eaten. Collared doves also take small amounts of fruit, including snow-berries that have been knocked to the ground by other birds. It seems that in certain regions of Europe collared doves are becoming pests of cherry orchards.

Until the early 1900s the collared dove was found mainly in Africa and Asia, but the species colonized much of Europe within just 60 years.

Rapid breeding

Generally collared doves nest in trees, preferably coniferous species. They build the nests near the trunk, about 6–60 feet (1.8–18 m) above ground. Nests are sometimes found on roofs and window sills, and in a few instances the abandoned nests of pigeons and other doves are used as foundations. The male collared dove helps the female with nest building, collecting twigs from the ground or snapping them off the branches with his bill and taking them to the female, which takes care of the construction herself.

During courtship there is a display flight in which the pair of doves soar up, clapping their wings together over their backs. The birds then glide downward, often in a spiral, with wings and tail spread, showing the conspicuous black and white of the tail feathers.

The female collared dove lays two white eggs, rarely three, and occasionally just one in clutches that are late in the season. The eggs are incubated for 14–18 days. Both sexes share in the incubation, and, as is usual in doves and pigeons, the male sits during the day and the female by night. In India, observations revealed that the female sat for 18 hours at night.

Collared dove chicks spend 2–3 weeks in the nest and are fed for the first few days on milk secreted from the adults' crop. Dove milk is a cheesy fluid, very rich in proteins and fat, and to obtain it the chicks push their heads down the parents' throats. When they are a few days old, the milk is supplemented with grain and other seeds. A suggested function of dove milk is to supply vital proteins, which are almost lacking in a seed-based diet. Doves do not provide their chicks with insects, which other seed-eating birds collect to give their offspring the protein necessary for rapid growth. In northern Europe collared doves usually breed from March to September. When the young doves fledge they form flocks, which are swelled by birds from later broods. On average, only one chick survives from each brood, but this still allows for a rapid increase in numbers.

Due to its short incubation and fledging periods the collared dove can raise up to six broods each year.

COLLARED DOVE

CLASS **Aves**

ORDER **Columbiformes**

FAMILY **Columbidae**

GENUS AND SPECIES *Streptopelia decaocto*

ALTERNATIVE NAMES
Eurasian collared dove; collared turtle dove

WEIGHT
6–8½ oz. (170–240 g)

LENGTH
**Head to tail: 12–13 in. (30–33 cm);
wingspan: 18½–22 in. (47–55 cm)**

DISTINCTIVE FEATURES
**Pale, gray-buff plumage, with black
half-collar on hind neck; square-ended tail**

DIET
Seeds of grasses, including cereals; some fruits

BREEDING
**Breeding season: March to November;
number of eggs: usually 2; incubation period:
14–18 days; fledging period: 15–19 days;
breeding interval: 3 to 6 broods per year**

LIFE SPAN
Up to 13 years

HABITAT
Cultivated land, gardens and semiarid country

DISTRIBUTION
**Much of Europe and parts of southern Asia;
introduced to southeastern U.S.**

STATUS
Very common

Collared dove

COLOBUS MONKEY

LITTLE IS KNOWN OF the several species of colobus monkey because they are rarely seen, live in dense foliage and are generally uncommon or rare within their respective ranges. The best-known species are the five black and white colobus monkeys of the genus *Colobus*, which have short, black fur with long plumes of white running down the sides and on the tail. They also have brilliant white flashes on the chin, cheeks and forehead. The largest group of colobus monkeys, the red colobus monkeys of the genus *Piliocolobus*, consists of eight mainly redbrown species with black, gray and white areas of varying size and shape. The olive colobus, *Procolobus verus*, has dull grayish underparts and an olive green upperside that becomes brown on the back. The classification of the colobus monkeys has been much debated; some authorities place all of the species in a single genus.

Colobus monkeys (colobids) are mediumsized with long limbs and a rather small head in proportion to the body; their thumbs are absent or very small. The black colobus (*C. satanas*),

guereza (*C. guereza*) and Angola pied colobus (*C. angolensis*) measure about 3–5 feet (90–150 cm). Over half of this length is made up of the tail, which the monkey uses to balance. The olive colobus is the smallest species: its head and body measure 1¾–1⅔ feet (43–50 cm).

Shy treetop-dwellers

Colobus monkeys live in Africa from Senegal east to western Ethiopia and south to southern Congo. They inhabit dense forests and rarely come down from the trees. Olive colobus monkeys occasionally forage on the ground, but usually descend only to visit saltlicks (natural salt deposits). Colobus monkeys live in family groups of up to about 20 individuals, led by an old male. Each group has a territory, the boundaries of which are defended by threatening calls. When they sense danger, the monkeys' natural reaction is to hide. They are very difficult to see in the forest as the plumes and light and dark patterns of some species blend in well with the sun-dappled forest vegetation.

Red colobus monkeys favor moist, evergreen forests situated close to water. Most species are found in lowlands.

One of the few colobus monkeys that is still relatively plentiful in much of its range, the guereza colobus has a thick tail that may reach 35 inches (90 cm) in length.

COLOBUS MONKEYS

CLASS	**Mammalia**
ORDER	**Primates**
FAMILY	**Cercopithecidae**

GENUS AND SPECIES **Black and white (pied) colobus: 5 species, including black colobus,** *Colobus satanus*; **Angola pied colobus,** *C. angolensis*; **and guereza colobus,** *C. guereza*. **Red colobus: 8 species, including western red colobus,** *Piliocolobus badius*; **and Central African red colobus,** *P. oustaleti*. **Olive colobus: 1 species,** *Procolobus verus*.

WEIGHT
6½–33 lb. (3–15 kg)

LENGTH
Head and body: 16–30 in. (40–75 cm); tail: 20–40 in. (50–100 cm)

DISTINCTIVE FEATURES
Small head; very long tail; thumbs absent. *C. satanus*: **all black; other** *Colobus*: **mainly black with flashes of white.** *Piliocolobus*: **red and brown with variable black, white and gray patches.** *Procolobus*: **dull greenish gray.**

DIET
Leaves, shoots, fruits, seeds and flowers

BREEDING
Age at first breeding: 3–6 years (male), 2–4 years (female); breeding season: all year; number of young: 1; gestation period: 120–180 days; breeding interval: variable

LIFE SPAN
Up to 20 years

HABITAT
Thick forest and rain forest; mainly lowlands

DISTRIBUTION
West and Central Africa

STATUS
All species declining; 2 species critically endangered; 5 species endangered

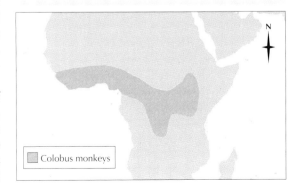

Colobus monkeys

Normally a group of colobus monkeys stays near the center of its territory. The members of one group that was studied had a 30-acre (12-ha) territory but tended to keep to a central 9-acre (3.5-ha) area where they followed regular tracks. They set off in search of food at sunrise, returning to the trees just before sunset to sleep.

Colobus monkeys eat mainly the foliage of forest trees. Some trees are preferred over others, and young, tender leaves in particular are selected. Digestion of coarse leaves, which are not very nutritious, is improved by the colobus monkeys' complex stomach, which is divided into pouches, rather like the stomachs of cattle. The stomach holds one-third of a monkey's body weight in food, and the animal must take long rests between eating for digestion. Colobids also feed on other plant materials that are hard to digest, including unripe fruits, seeds, seed pods and petioles (leaf stalks). The lengthy periods of rest between meals enable bacteria in the stomach to ferment and break down plant cellulose and to detoxify compounds in seeds and leaves.

Humans and eagles are the colobus monkeys' main enemies, and they are also hunted by chimpanzees. When threatened a group may abuse the intruder, shaking branches and using a wide range of aggressive gestures and calls. Otherwise the colobus monkeys try to hide or flee, the leader bringing up the rear.

Breeding behavior

A pronounced hierarchy exists among groups of colobus monkeys, and most matings are performed by the dominant male. Shortly before a female gives birth she leaves the group. The newborn baby, which is white all over, is held in the mother's arms at first, but when it is 2 weeks old it can cling to her back as she climbs through the trees with the rest of the group. At 6 weeks the infant begins to eat leaves, but it is not fully weaned until 7 months old, when it leaves its mother and joins the other youngsters.

The olive colobus is most unusual among monkeys in that the female carries her newborn baby in her mouth. Only several weeks later does the youngster cling to her fur. This may be due to the shortness of the mother's fur, which prevents a very young baby getting a grip.

Under threat

The loss of forest habitats and overhunting has seriously damaged many populations of colobus monkeys and today most species are regarded as being at risk. The I.U.C.N. (World Conservation Union) classifies five species as endangered: the olive and black colobus monkeys and three species of red colobus. Two species with limited distributions are regarded as critically endangered: the Zanzibar red colobus, *Piliocolobus kirkii*, and the Tana River red colobus, *P. rufomitratus*. The latter is threatened by deforestation to provide land for agriculture and by dramatic changes in the surviving vegetation due to dam construction, water diversion and forest fires.

Colobus monkeys have also suffered historical declines. Black and white colobus fur was used by certain African tribes for ceremonial capes, headdresses and shields, and in the past Arab traders carried colobus furs to central Asia where the naturally glossy black fur with white plumes was valued highly. From Asia samples found their way to the furriers of Venice who were reputed to have believed that craftspeople had skillfully fixed white plumes into the black fur. In the 19th century, the fur became popular in Europe and by 1892 175,000 skins were exported there; the market's rapid growth was made possible by hunting with modern firearms.

The guereza colobus lives in small territories in groups consisting of related offspring and a few hierarchically ranked males.

COLORADO BEETLE

After overwintering in the soil, adult Colorado beetles emerge in the spring to feed and mate. The eggs are usually laid on the underside of potato leaves and hatch a few days later.

THE COLORADO BEETLE is the most destructive insect defoliator of potatoes. It also causes significant damage to other members of the plant family Solanaceae, including cultivated species such as tomatoes, peppers, red peppers, tobacco and eggplant (aubergine). Among the beetle's other host plants are several wild species of the genus *Solanum*, including black nightshade and woody nightshade.

The adult Colorado beetle is about ⅖ inches (1 cm) long, a little bigger than a ladybug. Its convex, shiny back is longitudinally striped with black and yellow, and the thorax, the region just behind the head, is spotted black and yellow. The beetle's Latin species name, *decemlineata*, means "ten-striped," as there are five black longitudinal stripes on each wing cover. The larva is orange-yellow with black markings on the head, black legs and three rows of black spots along each side of the body. It has a characteristic hump-backed appearance.

Life cycle of a potato pest

The Colorado beetle is a serious agricultural pest, feeding on potato leaves both as a larva and an adult. It passes the winter as a mature beetle, hibernating underground in the soil at a typical depth of 10–12 inches (25–30 cm). One of the beetle's most commonly used overwintering sites is woodland adjacent to the fields in which it spent the previous summer. In the late spring the beetle emerges and, if it does not find itself in an area rich with potato plants, immediately flies in search of them, frequently for a distance of many kilometers.

The adult female Colorado beetle lays her eggs on the leaves of the selected plant, usually in clusters on the underside of each leaf. The eggs are yellow in color, about 1.5 mm in length and ovoid in shape. The larvae hatch in a few days, the hatching period depending on the temperature, and start to feed voraciously on the leaves of the host plant. They are fully grown in about 3 weeks.

The larvae then burrow into the soil to pupate, and a new generation of adult beetles emerges in 10–15 days. If the environmental conditions remain suitable a third generation of adults may be produced. The Colorado beetle is tolerant of climatic extremes, from desert to near-Arctic conditions. As soon as bad weather sets in, however, the beetles burrow into the soil and hibernate until the spring.

Colorado beetles damage the haulm, the part of the potato plant that grows above ground. A plant may be stripped of its leaves by both larvae and adults, which means that the tubers cannot develop. The large number of eggs produced by each adult female and the rapid succession of generations make the Colorado beetle a formidable pest. A single individual emerging in the spring may have given rise to thousands of descendants by the fall.

Insect predators

About 38 species of insect worldwide are known to feed on Colorado beetles, including bugs, beetles, wasps and flies. Some of these species are successful in acting as biocontrol agents and in reducing the spread of the Colorado beetle. In tests, heavy releases of the predacious bugs *Perillus bioculatus* and *Podisus maculiventris* were shown to suppress beetle density by 60 percent, with a subsequent reduction in defoliation of 85 percent and a 65 percent increase in potato yield. The parasitic wasp *Edovum puttleri* lays its eggs within those of the Colorado beetle; it has been

COLORADO BEETLE

PHYLUM **Arthropoda**

CLASS **Insecta**

ORDER **Coleoptera**

FAMILY **Chrysomelidae**

GENUS AND SPECIES **Leptinotarsa decemlineata**

ALTERNATIVE NAMES
Colorado potato beetle; potato bug

LENGTH
About ⁴/₁₀ in. (1 cm)

DISTINCTIVE FEATURES
Adult: convex, oval-shaped abdomen; 10 black longitudinal stripes on yellow elytra (wing cases); brownish red head. Larva: hump-backed body; orange-yellow overall with 3 rows of black spots along each side.

DIET
Plants of family Solanaceae, such as woody nightshade and black nightshade; cultivated crops, including potatoes, tomatoes, eggplants, tobacco, red peppers and peppers

BREEDING
Age at first breeding: 4–5 weeks; breeding season: spring and summer; number of eggs: 300 to 500 laid within 5–6 days of mating; hatching period: up to 7 days; larval period: usually about 3 weeks

LIFE SPAN
Up to 1 year

HABITAT
Any environment suitable for growth of potatoes and related plants in the family Solanaceae; can survive extremes of climate

DISTRIBUTION
Native to southwestern U.S. and Mexico but now also occurs in 3 million sq mi. (8 million sq km) of North America and 2 million sq mi. (6 million sq km) of Eurasia

STATUS
Globally abundant

found to parasitize 80–90 percent of potato beetle egg masses on eggplants, killing up to 80 percent of the beetle eggs per mass.

Spraying potato foliage with an arsenical compound is the method usually employed to control the Colorado beetle. However, although a variety of chemicals has been used to control the pest, including organochlorines, carbamates, organophosphates and pyrethroids, the species consistently develops an immunity to the chemicals used against it.

An entomological curiosity

Almost all species of insect are conditioned to live in a particular type of climate. If the prevailing climate is not suitable the insects will fail and die out at some stage in their life cycle. The Colorado beetle is a conspicuous exception to this rule. It can live all year round in Canada, where the winters are Arctic in severity, and also survives in the hot deserts of Texas and Mexico. The beetle's habit of hibernating deep underground as an adult, from the end of August or the beginning of September, is probably the most important factor enabling it to adapt to any climate in which potatoes are grown.

A potato bridge

In the early 1820s, the American explorer Stephen Harriman Long discovered a black-and-yellow-striped beetle in the Rocky Mountains of Colorado state. The insect has subsequently become known as the Colorado beetle. In the wild the beetle fed on a type of nightshade called buffalo burr, *Solanum rostratum*, a member of the potato family. The potato is native to Peru and Ecuador, in South America, but was brought to Europe by the Spaniards and later found its way to the new colony of Virginia in North America. In the course of the opening up and settlement of western North America in the 1850s, potatoes were introduced and cultivated by the pioneers. By 1859 the Colorado beetle was discovered to be feeding on cultivated potatoes in Nebraska.

Colorado beetle numbers increased rapidly and the species began to spread across the continent. No control measures were in effect at that time and the beetle spread quickly from potato field to potato field, frequently destroying whole crops. From Nebraska in 1859 it appeared in Illinois in 1864 and in Ohio in 1869, reaching the Atlantic coast in 1874, an average rate of travel of 85 miles (135 km) a year. The potato fields of the United States had formed a bridge from west to east along which the beetle could travel. It also spread northward into Canada.

The Atlantic formed a barrier to further progress until 1922, when the beetle was found in the Gironde region of France. From this area it has extended its range all over continental Europe. The Colorado beetle has appeared in western China and Iran and is still expanding its range. Among the regions potentially at risk from this pest are Korea, Japan, Russian Siberia, areas of the Indian subcontinent, parts of North Africa and the temperate Southern Hemisphere.

COMMON DOLPHIN

Common dolphins live in groups, called schools, composed of individuals of both sexes and all ages.

SMALLER THAN THE BOTTLENOSE dolphin, *Tursiops truncatus,* the common dolphin is up to 8 feet (2.5 m) long and weighs up to 245 pounds (110 kg). Males are slightly larger than females. The beak is narrow and sharply cut off from the forehead, and there are 40 to 44 teeth on each side of the upper and lower jaws. The common dolphin is mainly black or dark brown on the back and flanks, sometimes with brown or violet and light spots, and creamy white below. A distinctive tan-colored hourglass pattern is distinguishable on the flanks, giving rise to one of the species' alternative names, the hourglass dolphin. A dark stripe runs from eye to snout. The coloration of the area near to the sexual organs is an indicator of a dolphin's sex: the males have a prominent black stripe just above the sexual opening; females have a somewhat thinner stripe, with gray countershading.

The common dolphin's pale underbelly may serve to camouflage it from below against the brighter light of the upper levels of the ocean. Its mixture of pale blazes and dark stripes may also act to break up the contours of its body in the water, thereby making the dolphin less noticeable to both predators and prey. Two major structural variations are recognized, the short-beaked and long-beaked forms, though these are not considered to be sufficiently different to be separate species. The short-beaked form is generally found offshore, while the long-beaked form is most frequently found inshore.

There are several other species in the same genus as the common dolphin. *Delphinus* dolphins occur between New Guinea and Australia, from South Africa across the Indian Ocean and throughout the West Pacific to Japan.

Worldwide distribution

Common dolphins inhabit most tropical and temperate seas. They are found in the Pacific, as far north as Japan and northern California, and south to southern Australia, New Zealand and southern Chile. In the Atlantic they never go farther north than Iceland and Finnmark (northern Norway). Common dolphins do not generally occur as far north or south as bottlenose dolphins. Some populations of common dolphins appear to be migratory: in summer schools of the species can be seen in Algerian waters in the western Mediterranean, but they disappear in winter. Scientists believe that, as

COMMON DOLPHIN

CLASS	**Mammalia**
ORDER	**Cetacea**
FAMILY	**Delphinidae**
GENUS AND SPECIES	***Delphinus delphis***

ALTERNATIVE NAMES
White-bellied porpoise; saddleback dolphin; crisscross dolphin; hourglass dolphin

WEIGHT
155–245 lb. (70–110 kg)

LENGTH
Head and body: 6–8 ft. (1.8–2.5 m)

DISTINCTIVE FEATURES
Recurved, pointed dorsal fin; pointed flippers (pectoral fins); narrow beak; gray or tan hourglass pattern along side of body; creamy white chest; black line from flipper to lower jaw and from eye to snout; black or brown-black upper flanks and back. Two forms, distinguished by beak length.

DIET
Fish, squid and cuttlefish

BREEDING
Age at first breeding: about 8 years; breeding season: varies according to region; number of young: 1; gestation period: about 270–300 days; breeding interval: 16–28 months

LIFE SPAN
Probably up to 25 years

HABITAT
Open ocean and inshore waters

DISTRIBUTION
Tropical and temperate seas worldwide; does not range as far north and south as bottlenose dolphin, *Tursiops truncatus*

STATUS
Very common; localized declines in Pacific

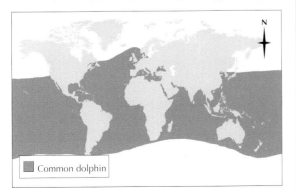

Common dolphin

with other species, migration in the common dolphin may be related to the seasonal movements of its main prey.

Many aspects of the common dolphin's habits, life history and physiology are close to those of the bottlenose dolphin. Their diving and swimming mechanisms are very much alike, though common dolphins cannot stay submerged as long as bottlenose dolphins. In the former species the average duration of a dive is 2–3 minutes and it is reported to die if kept under for 5 minutes. Dolphins breathe through a blowhole in the top of the head; when they move underwater this hole is sealed off by a muscular flap, to prevent them from inhaling water.

Common dolphins are among the fastest members of the order Cetacea. While they may cruise at about 5 knots (10 km/h), they can swim at 20 knots (40 km/h) for a considerable length of time. Faster speeds have been recorded with dolphins swimming in ships' bow-waves, when they hold their bodies at an angle, using them in the manner of surfboards. Dolphin skin is flexible and ripples as the animals move through water, thus reducing drag.

Complex social structure

Common dolphins travel in schools which, as in bottlenose dolphins, are made up of both sexes and all ages. There is no leader but males bear

It is possible that common dolphins leap out of the water in order to dislodge external parasites or to signal to others.

Schools of common dolphins are broader than they are long. Such a formation enables the dolphins acoustically to scan a greater area than would otherwise be possible.

the scars of fights with one another. The schools may be very large, and in the Black Sea a school of 10,000 was once reported. Such large schools must be rare, however, and probably form only where there are even larger concentrations of prey fish. The size of the basic social unit in common dolphins may be less than 30.

Common dolphins produce the full range of sounds made by most delphinids, including creaks, whistles, clicks and squeaks. Dolphins use high-frequency clicks as a form of echolocation or radar: the sounds emitted bounce off surrounding surfaces and enable the dolphins to identify objects and to navigate. When moving in schools of 100 or more, whistling becomes almost synchronous and seems to be timed with each layer of dolphins leaping from the water. It is believed that these whistles act as a form of communication.

Playing schools

It is not unusual to see a school of dolphins or porpoises playing, and the common dolphin seems to be the most playful species of all. A school can be watched from a ship or cliff for an hour or more without the dolphins traveling out of sight. This behavior has often been considered an instance of play for its own sake, rather than

for a strictly useful purpose, as is the case with nearly all forms of animal behavior. When dolphins swim normally, they follow an undulating path, continually rising to come to the surface to breathe. This behavior is known as transitting. Usually only the top of the head breaks the surface of the water, but in the simplest form of play the dolphins exaggerate the rise and fall so that they leap out of the water, curving over in an arc to fall back in again. Common dolphins are also often observed to leap out vertically and land on their back, belly or side, to roll over and over or to swim on their backs. Occasionally they swim at the surface, raising the tail and bringing it down on the water with a loud smack every 2 or 3 seconds.

Many experts believe that this behavior may not be just play for its own sake. It is possible that such movements may help to dislodge external parasites, act as a signal to other dolphins, or serve some other social function.

Breeding and feeding

The calves are born from midwinter to summer after a gestation period of 9 months, although the times of conception and birth vary geographically. In the north Pacific, April–June and October–December are generally regarded as the peak periods for conception, with March–April and September–October as the periods when most births take place. In the North Atlantic, July–October is the peak period for conception, and June–September is the peak birth period.

The breeding habits of the common dolphin resemble those of the bottlenose dolphin. The calf is born tailfirst and the mother helps it to the surface to take its first breath. If it is injured she will support it until it dies. Suckling takes place underwater and milk is forced into the calf's mouth by the contraction of muscles around the mammary gland. This lactation period generally lasts for about 19 months.

The diet of the common dolphin consists of fish, squid and cuttlefish, and varies according to the season and location. Flying fish and other species that live near the surface are also taken. Common dolphins off southern California feed on squid and anchovy during winter, but apparently switch to smelt and lantern fish during the spring and summer months. Dolphins consume the food equivalent of nearly one-third of their body weight each day.

Conservation

Certain forms of fishing, particularly those practised in the Pacific, often result in the incidental capture of common dolphins. Fisheries based in the eastern tropical Pacific use purse-seine nets to capture yellowfin tuna. This fishing technique utilizes a large seine, or net, which is suspended between two fishing boats to surround entire shoals of fish. The ends of the net are drawn together, thus enclosing marine animals indiscriminately. The animals within, which regularly include dolphins of various species, are trapped and eventually drown. This procedure has provoked international controversy: it is certainly the most economically viable method of gathering tuna, but ecologically it is extremely damaging. It is estimated that more than 6 million dolphins have been killed in purse-seine nets since this form of fishing was introduced.

Common dolphins are also illegally hunted for sport as well as for their meat, while overfishing depletes the dolphins' food resources. Pollution, such as toxic waste washed out to sea, reduces the animals' immunity to disease and affects their reproductive abilities. All of these issues have caused widespread international concern, although moves to adopt more dolphin-friendly fishing methods have to be weighed up against the economic needs of individual countries. Dolphins have been fully protected in Australian waters since the 1980s and dolphin deaths caused by U.S. vessels are now believed to stand at less than 1,000 per year.

Common dolphins remain abundant in global terms but have suffered large declines in the Pacific, due mainly to overfishing.

COMMON FROG

strictly accurate. They prefer shallow water a few inches deep, but will spawn in water over 3 feet (90 cm) deep if necessary.

Certain ponds contain spawn every year while others nearby never do. Why some should be preferred to others is not clear, but an individual frog will return each year to the same pond, and newly built-up areas are sometimes invaded by frogs looking for ponds that have been filled in. Together with the infilling of ponds by farmers, this has probably contributed to the decline of frogs in some areas.

Outside the breeding season common frogs lead a solitary life on land, usually in damp places near ponds and marshes, but they can also be found some distance from water squatting in grass. In the fall, the exact time depending on the weather, common frogs enter hibernation, or rather torpor, for they will become active on mild days. The date of emergence from hibernation is also weather-dependent.

Spawning

Shortly after hibernation, common frogs return to fresh water. It has been suggested that they locate water by the smell of the algae growing there, which is essential for the growth of the tadpoles (larvae). However, the arrival of frogs at filled-in ponds seems to disprove this, and there is evidence that frogs and toads navigate by the sun or stars. The male common frogs arrive first, the females soon afterwards. They may wait a week before spawning, but breeding can be over in a couple of days.

Male common frogs croak, mainly by day, but are very quiet compared with some species of frogs and toads. Croaking is produced by passing air backward and forward across the vocal chords from lungs to mouth. The mouth is not opened, so common frogs can croak while under water. During croaking, the throat is distended under the chin.

For the duration of the spawning period the males will attempt to embrace any object, wrapping the front legs around it and holding on with a vicelike grip, aided by the special swellings that develop on their hands at this time. This mating embrace is called amplexus. If one male clasps another, the latter grunts to show the former its mistake. Males can also recognize gravid (pregnant) females by their size and by the newly rough texture of their skin. Mating lasts about a day, and a male may mate with

The common frog has a smooth skin with many small, wartlike slime glands that help keep its skin moist when out of water.

THE COMMON FROG OF EUROPE is highly variable in color. It is usually green or brown, but alternatively its ground color may be gray, yellow, orange or even red. Certain markings, however, are consistent. There are dark transverse bars on the hind legs, a streak on the forearm and in front of the eye and a brown patch behind each eye. Other marks include speckles and marbling. The underparts are lighter, dirty white in males and yellow in females.

The common frog's skin is smooth with many small, wartlike slime, or mucous, glands. These keep the skin moist while the frog is not in water, preventing excessive water loss. In the breeding season the males' skin becomes more slimy and the warts on the females become larger and pearly white. Another amphibian, the common toad, *Bufo bufo*, resembles the common frog but has dull, wrinkled skin and shorter legs.

Distribution

Common frogs range over most of Europe as far east as the Ural mountain range. The species is usually seen only in the breeding season when large numbers gather in ponds to mate and spawn. The assertion that frogs do not discriminate in their choice of a pond for spawning is not several females. Each female lays 1,000 to 2,000 eggs, which are ejected within 5 seconds. The male immediately sheds his sperm over them, then the pair disengage. The eggs are 2.5 millimeters in diameter at first and sink to the bottom of the water, but the jelly surrounding them

COMMON FROG

CLASS	**Amphibia**
ORDER	**Anura**
FAMILY	**Ranidae**
GENUS AND SPECIES	***Rana temporaria***

ALTERNATIVE NAME
Grass frog

LENGTH
Up to 4 in. (10 cm); female larger than male

DISTINCTIVE FEATURES
Bright, smooth skin; color very variable, usually green or brown with dark markings but may be grayish, yellowish or reddish

DIET
Adult: insects, slugs, snails, worms, spiders, mites and centipedes. Larva (tadpole): algae, amoebae, bacteria and small crustaceans.

BREEDING
Age at first breeding: 3 years; breeding season: early spring; number of eggs: 1,000 to 2,000; hatching period: 14 days; larval period: 3 months; breeding interval: 1 year

LIFE SPAN
Up to 12 years in captivity

HABITAT
Moist places on land; breeds in shallow freshwater ponds, ditches and marshes

DISTRIBUTION
Most of Europe (except Mediterranean) east to Ural Mountains

STATUS
Common, but declining in many areas

Common frog

swells up and they become buoyant, floating to the surface as masses of frog spawn. The jelly is not eaten by the developing tadpole but serves as protection against bacteria, fungi and the cold. When the surrounding water is freezing, frog spawn is a degree or so warmer because the black egg absorbs the radiant heat from the sun that is trapped by the jelly.

From tadpole to frog

In about 2 weeks the tadpoles have developed and the egg membrane is dissolved or digested by a special gland. After emerging each tadpole clings to the remains of the jelly mass with adhesive organs. At this stage it has external gills, no limbs and the head, body and tail are merged together. Over time, the external gills disappear, the mouth opens and the tadpole begins to feed, while the internal gills develop from buds on the sides of the head. The gills are covered by a flap of skin, the operculum. Later, the limbs grow, the hind legs appearing first as the forelegs are covered by the operculum. Finally, the gills are absorbed as the lungs start to function and the tail is absorbed into the body, resulting in a fully formed froglet after 3 months. The froglets move to the edge of the water body until their legs strengthen, enabling dispersal. In mountains tadpoles may overwinter, becoming froglets in spring. They are sexually mature in 3 years.

Adult common frogs prey mainly on slugs and snails, but also take insects, spiders, mites, centipedes and worms. They lie in wait and catch passing prey with their extensible tongues. Tadpoles feed on bacteria, amoebae, crustaceans such as daphnia and algae, which they scrape off water plants with their horny mandibles and lip teeth. Food and water are swept into a tadpole's gullet by currents set up by cilia and the food is then strained off, the water leaving through the gill slits and spiracle (breathing hole).

Common frogs mate in early spring, usually during March, after they have emerged from hibernation. The jellylike spawn (egg masses) hatches within 2 weeks.

COMMON TOAD

ESPITE A SUPERFICIAL resemblance to the common frog, *Rana temporaria*, few people have difficulty in distinguishing a common toad. It is squatter and more compact in appearance than the common frog, and has a flatter back and shorter legs. Instead of the moist, bright skin of the frog, the toad has a dull, wrinkled, warty skin. Its movements are slow and although it can jump a short distance, it usually walks over the ground. The common toad is more terrestrial than the common frog, although both species return to water to lay their eggs.

The common toad's rough skin blends well with the ground, so that it can easily be mistaken for a clod of earth. This impression is heightened by the dark brown or gray coloring, which can change, although only a little and slowly, to match the surroundings. The toad's jewel-like eyes are golden or coppery-red and behind them lie the bulges of the parotid (saliva) glands that produce an acrid, poisonous fluid. The common toad has no external vocal sac and, unlike many

Common toads are active mainly by night and during prolonged wet weather.

of its relatives, the male has a very weak croak. Adult males are 2½ inches (6 cm) long; females are about 1 inch (2.5 cm) longer, although they may reach 5 inches (13 cm).

The common toad is distributed throughout Europe, North Africa and northern and temperate Asia. It is found in much of the British Isles but is absent from Ireland.

Migrations

The common toad, like the common frog, hibernates from October to February, though it generally chooses drier places than that species. Hibernating toads are found in dry banks and the disused burrows of small mammals, and sometimes in cellars and outhouses. In the spring the common toad migrates to breeding pools, preferring deeper water than the common frog. Where the two species occur in the same ponds, the frog will often be in the shallows while the toad occupies the middle stretch of water.

The common toad gives the impression of being a much slower mover than most frogs. Its migration route soon becomes littered with the remains of toads that have been attacked by predators. Although the common toad's migration may be long and arduous, perhaps covering 2–3 miles (3–5 km) at a rate of ¾ mile (1.2 km) every 24 hours, the species is very persistent, and will climb stone walls and banks that lie in its way. It readily crosses roads, though many toads are run over by cars when doing so.

Outside the breeding season common toads live in hollows that they scoop out with their hind legs. In soft earth they bury themselves completely; in areas of firmer ground the hole is simply made under a log or stone. These homes are usually permanent, the toad returning to the same place day after day. Occasionally the retreats may be in places that require the toad to expend considerable physical effort; some specimens have even been found in bird nests.

Prey must be moving

At night and during wet weather, common toads emerge to feed on many kinds of invertebrate. The prey must be in motion for the toads to detect it, because the toads' eyes are adapted to react to moving objects. Any insect or other small invertebrate is taken, ants being particularly favored. Some less digestible animals such as burnet moth caterpillars or caterpillars covered with stiff hairs are left undisturbed, but toads are known to sit outside beehives in the evening and catch the workers as they come back to the hive.

COMMON TOAD

CLASS	**Amphibia**
ORDER	**Anura**
FAMILY	**Bufonidae**
GENUS AND SPECIES	***Bufo bufo***

LENGTH
Up to 5 in. (13 cm); female larger than male

DISTINCTIVE FEATURES
Dull, warty skin; relatively short legs; usually brownish or gray but can change color to match environment; eyes golden or reddish

DIET
Adult: mainly invertebrates such as ants, earthworms, slugs and snails; occasionally very small frogs, toads, newts and snakes. Larva (tadpole): algae and small crustaceans.

BREEDING
Age at first breeding: 4 years; breeding season: spring; number of eggs: 3,000 to 4,000; hatching period: 10 days; larval period: 3 months; breeding interval: 1 year

LIFE SPAN
Up to 40 years in captivity, usually much less

HABITAT
Cool, moist places on land; breeds in freshwater ponds and marshes

DISTRIBUTION
Europe east to Japan and south to North Africa, northern India and southern China

STATUS
Common

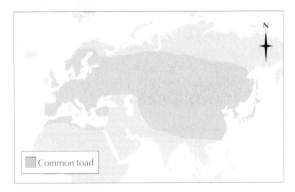
Common toad

Snails are crunched up and earthworms are pushed into the mouth by the toads' forefeet, which also scrape excess earth off the prey. Young newts, frogs, toads, slowworms and grass snakes are occasionally eaten. Toads often return to a favorite retreat after hunting and will use the same site for years.

Spawn in strings

There is little to distinguish the breeding habits of common toads and common frogs. They breed in spring at roughly the same time and may be seen in the same pools. Male common toads start arriving before the females but later in the spawning season may arrive already coupled with the females, riding on their backs. The spawn is laid in strings rather than in a mass. The eggs are embedded three or four deep in threads of jelly that may be up to 15 feet (4.5 m) long. Each female lays 3,000 to 4,000 eggs, which are smaller than those of the common frog, being less than 2 millimeters in diameter. The jelly swells up but the spawn does not float, because it is wrapped around the stems of water plants. The eggs hatch in 10 days and the tadpoles develop in the same manner as frog tadpoles, becoming shiny, black toadlets within about 3 months. Sexual maturity is reached in 4 years, before the toads are fully grown.

Defense mechanisms

Common toads have many predators, including herons, snakes, hedgehogs, foxes, gulls and crows. The poisonous secretions of their parotid glands are certainly effective against dogs, which salivate copiously after taking a toad in their mouth. Sometimes a common toad's reaction to a snake is to blow up its body in the manner of the burrowing toad, *Rhinophrynus dorsalis*. Older common toads may fall victim to the larvae of parasitic greenbottle flies, which crawl into their nostrils and eat their body tissues.

Male common toads fight aggressively for females. They have strong forearms to maintain a secure grip during the mating embrace, known as amplexus.

CONCH

THE WORD CONCH WAS originally applied to bivalve mollusks but was later taken to include mollusk shells in general, the study of which is still known as conchology. Today the word conch signifies only the sea-snails of the genus *Strombus*. Some other large marine snails are often called conchs, for instance the horse conch of the Florida coast, but these species do not belong to the genus of true conchs.

Large shells

Strombus shells range in length from just ¾ inch (2 cm) to the 13 inches (33 cm) of *S. goliath*. One of the best known species, the queen conch, *S. gigas*, which occurs throughout the Caribbean, reaches a length of about 12 inches (30 cm) and weighs up to 6 pounds (2.5 kg). This mollusk is sexually mature at 3 years, by which time it weighs 2 pounds (1 kg) and is 8 inches (20 cm) long, and continues to grow to its maximum size at the rate of 3 inches (7.5 cm) per year.

In some species of conch the lip of the shell is developed in the adult as a heavy projecting "wing" or fingerlike projection. This serves to stabilize the shell as it lies on the seabed, preventing it from rolling over. A peculiarity of conchs is the two large eyes, often ringed with orange, red or yellow, which are carried on long stalks arising from either side of a stout proboscis. Each stalk has a short tentacle, and the right eye is lodged, when in use, in the so-called stromboid notch in the lip of the shell.

Confined to warm water

Members of the genus *Strombus* are widespread in tropical waters, but are absent from most of the Atlantic. They can survive only in places where the water temperature never falls below 70° F (21° C). The distribution of conchs is therefore similar to that of coral reefs, and few species occur in waters too cool for reef corals.

Conchs are found mainly in the Indian Ocean and the South Pacific, with more than 40 species in the Red Sea, Arabian Sea and East Africa eastward through the Indian Ocean to Hawaii and Easter Island in the Pacific. There are seven conch species in the Caribbean, four on the Pacific side of Central America, and one on the West African coast. Two species live at depths of up to 400 feet (120 m), but most live in shallow water, from low-tide level to 65 feet (20 m).

Jumping shellfish

Conchs move in quite a remarkable way for marine mollusks. When at rest conchs tend to bury themselves in sand or gravel, but when active they push themselves along. The foot of a conch is unlike that of the more familiar terrestrial snails in that only a small part of it is used as a creeping sole in locomotion. The hind end of the foot has a horny plate called the operculum, which in many land snails serves to seal the entrance to the shell when the animal withdraws inside. In conchs the operculum is clawlike, with a sharp, often serrated edge. It is used not only in

Conchs gather in large numbers to spawn and also perform mass migrations when food is scarce.

CONCHS

PHYLUM	**Mollusca**
CLASS	**Gastropoda**
ORDER	**Mesogastropoda**
FAMILY	**Strombidae**
GENUS	***Strombus***
SPECIES	**Over 50, including *S. goliath* and queen conch, *S. gigas***

ALTERNATIVE NAMES
Fighting conch; trumpet shell; stromb shell

WEIGHT
Up to 6 lb. (2.5 kg); male smaller than female

LENGTH
2–12 in. (5–30 cm)

DISTINCTIVE FEATURES
Thick shell, often with a "wing" or fingerlike projections for stability; 2 large eyes on long stalks; right eye stalk protrudes through stromboid notch in outer lip of shell

DIET
Mainly seaweed and algae

BREEDING
Age at first breeding: 2–3 years; number of eggs: tens of thousands to hundreds of thousands; hatching period: 14–21 days; larval period: 14–21 days

LIFE SPAN
Adult: usually 5–6 years

HABITAT
Warm, shallow seas, mainly 35–65 ft. (10–20 m) deep; usually on sand, mud, coral rubble and meadows of seagrass and algae

DISTRIBUTION
Tropical seas that always exceed 70° F (21° C); absent from most of Atlantic

STATUS
Generally common; localized declines due to overfishing

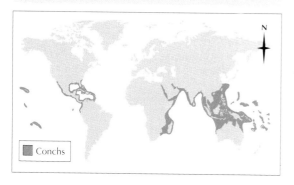

Conchs

defense, as a kind of dagger against crabs and fish, but also to pole the animal along with a leaping action. Each jump carries the conch forward by about half the length of its own shell. The conch pushes the operculum, which looks like a huge fingernail, into the sand or against a firm object and then presses hard.

The shell of a conch is nearly four times the weight of its soft parts and the thrust given by its muscles must be powerful. A conch shell 6 inches (15 cm) long will be carried forward 3 inches (7.5 cm) and lifted 2 inches (5 cm) off the seabed in the process. More expressive of the power of the thrust is the experience of conch divers who gather the shells to sell for food and as tourist souvenirs. The divers often arrive at the surface cut by the sharp edges of the conchs' opercula.

Seaweed grazers

The smaller conchs graze the deposits of seaweed fragments that accumulate on the seabed. Other conchs eat living seaweed. The queen conch, for instance, aligns its vertical slitlike mouth with seaweed fronds, seizing them in its two ribbed lips. Inside the lips is a pair of jaws as well as the usual radula, or horny tongue.

During the warmer months conchs gather in the shallows for spawning. After mating the females lay thousands of eggs in long, jellylike masses, up to 75 feet (23 m) long. The eggs hatch in 1–2 weeks, releasing free-swimming larvae that feed on plankton for up to 10 weeks before settling on the seabed.

The queen conch lives in the Gulf of Mexico and the Caribbean. Its empty shells are often washed up onto the region's beaches.

CONDOR

Despite its size the Andean condor feeds almost entirely on carrion. It rarely kills animals, and such live victims are invariably dying or wounded.

THE TWO LARGEST FLYING birds in the world, the Andean condor, *Vultur gryphus*, and California condor, *Gymnogyps californianus*, have wingspans that can reach just over 10 feet (3 m) in length. This is less than that of the wandering albatross, *Diomedea exulans*, a seabird found in the oceans of the Southern Hemisphere, but the condors' broad wings give them a much larger wing area. They weigh 20–25 pounds (9–11.5 kg) on average, as much as a large turkey, but would be dwarfed by their extinct relative *Teratornis incredibilis* of North America, which had a wingspan of about 16 feet (5 m).

The condors belong to the family Cathartidae, which also contains the New World vultures, and like those species are unrelated to the true vultures of the Old World except in appearance and habit. The adult California condor has naked pink skin on the head and its eyes and ears stand out behind a powerful hooked bill. Around its neck is a ruff of blackish plumes that matches its body plumage. The Andean condor, which is the larger of the two species, is darker around the head and in the adult male carries a fleshy comb (crest) and wrinkles hanging from the neck. Around its neck is a

brilliant white ruff; the remaining plumage is black, except for prominent silver-gray patches on the wings. Young Andean condors have a brownish tinge to much of their plumage.

Riding the air currents

A condor may fly over many hundreds of square miles in search of food. This it does almost effortlessly despite its great weight because, like an ocean-going albatross, it is adapted to make the most efficient use of air currents. Condors use rising bubbles of warm air known as thermals to gain lift. Thermals are caused by layers of air heating up near the ground and rising through the atmosphere. Glider pilots use thermals in the same way as condors, gliding in a series of circles within the thermal and using the rising air to lift them upward.

Andean condors have been recorded at altitudes of up to 15,000 feet (4,500 m), soaring up and over thunderstorms. Flying at such heights is exceptional, however, and condors usually rise no more than a few thousand feet before leaving one thermal and gliding away to another. In this way the birds are able to cover their home range with hardly a wingbeat.

ANDEAN CONDOR

CLASS	**Aves**
ORDER	**Falconiformes**
FAMILY	**Cathartidae**
GENUS AND SPECIES	***Vultur gryphus***

WEIGHT
17–33 lb. (8–15 kg)

LENGTH
**Head to tail: 3–4½ ft. (0.9–1.3 m);
wingspan: up to 10½ ft. (3.2 m)**

DISTINCTIVE FEATURES
**Huge size; massive hooked bill; very broad,
long wings; black body with gray areas on
wings and white neck ruff; naked head and
upper neck. Male: brown comb (on forehead)
and wattles (hanging from neck). Female:
black head. Juvenile: brownish plumage.**

DIET
**Carrion from large mammal carcasses;
mammal afterbirths; on coast takes dead
fish, seals and whales and bird eggs**

BREEDING
**Age at first breeding: 6–7 years; breeding
season: eggs laid February–June (Peru),
September–October (Chile); number of
eggs: 1; incubation period: 60 days; fledging
period: 180 days; breeding interval: 2 years**

LIFE SPAN
Up to 50 years

HABITAT
**Open grasslands in high mountain ranges;
visits coastal deserts and cliffs in Chile and
Peru and lowland grasslands in Argentina**

DISTRIBUTION
**South America in Andes Mountains, Tierra
del Fuego and neighboring lowland regions**

STATUS
Generally uncommon; rare in parts of range

Andean condor

Carrion eaters

In common with both Old and New World
vultures, condors are mainly scavengers. The
naked head and upper neck of the California and
Andean condors are adaptations for carrion-
eating, as feathers on the head would become
matted with blood and gore from thrusting
inside a carcass. Both species have powerful
hooked bills that are strong enough to tear
through animal hide and muscle and yet agile
enough to nibble delicately at bones. The Andean
condor's diet consists largely of carcasses of the
llama and guanaco, relatives of camels.

Although carrion is the main food of the
Andean condor on rare occasions it takes weak
or dying lambs, llamas and deer. The species also
descends to the coasts of Chile and Peru to feed
on dead fish, seals and whales and to scavenge
eggs at seabird colonies. Mammalian afterbirths
are highly nutritious and taken when available.

Slow reproductive rate

Condors breed extremely slowly and are therefore very vulnerable to hunting pressures and to
habitat changes because their populations take a
long time to recover from declines. Condors do
not mature sexually until 6 or 7 years of age, and
although they may live for 50 years or more, lay
only one egg every other year. Courtship in
condors consists of a dance with the male and
female walking backward and forward, hissing
and clucking with wings spread. The birds circle
one another, then push and peck vigorously, and
they also perform aerial chases. Breeding pairs of
condors mate for life.

The single egg is laid on the bare rock of a
ledge or cave high up on a cliff face. Incubation
lasts 7–9 weeks, and both sexes participate in

*The Andean condor
(below) has a massive
wingspan of up to
10 feet (3 m) or more,
the largest of any
landbird. The much
rarer California condor
is slightly smaller.*

that condors would land and even attack a model animal, but could not find a rotting carcass hidden from sight by a tarpaulin. The three New World vultures of the genus *Cathartes*, the turkey vulture (*C. aura*), lesser yellow-headed vulture (*C. burrovianus*) and greater yellow-headed vulture (*C. melambrotos*), are exceptions: they can smell carcasses hidden in thick forest.

It seems that sight is the sense that condors use to detect food. The birds of prey of the order Falconiformes typically have very keen eyesight. Their survival depends on locating food from far away; they are skilled in detecting carcasses, prey or others of their kind, often at great distances.

The California condor

One of the rarest birds in the world, the California condor is now classed by the I.U.C.N. (World Conservation Union) as critically endangered. At one time the species was relatively common, however, and its range extended from Oregon to California, east to Nevada and New Mexico. In prehistoric times the condor was more widespread: fossil recoveries show that it occurred across North America including as far east as Florida and New York State.

When European settlers started to filter into the western highlands of the United States the California condor was regularly seen in flocks of 20 or more. However, subsequent human settlement destroyed its habitat and greatly reduced the numbers of large mammals, the carcasses of which it depended upon. The California condor was often shot because it was a large target and was thought to spread disease. Many condors died after eating poisoned carcasses left for wolves. More recently the species has suffered from collisions with power lines and other artificial objects.

It was estimated that there were only 60 California condors left in the wild in 1953, and this number had fallen to just 21 by 1982. In 1987 it was agreed that all wild California condors should be captured for a captive-breeding program; the species had been deliberately made extinct in the wild. The first young condors were released back into the wild in 1992. By the late 1990s the total population of California condors stood at about 95 captive and 55 released birds. There are currently two release sites: Los Padres National Forest, near Los Angeles in California, and the Colorado River, Arizona; it is also planned to release birds in the Grand Canyon.

Decades of decline reduced the California condor's population to less than 30 birds by the 1980s. A complex and very expensive conservation program increased this figure to about 150 birds by the late 1990s.

keeping the egg warm. The young condor stays with its parents for at least a year and for several years it can be distinguished from adult birds by the down covering its head and neck.

Finding food

Charles Darwin, Alexander von Humboldt and other early European visitors to South America were impressed by the Andean condor's majestic flight and by its ability to soar until lost from sight. But zoologists were puzzled by the birds' ability to find food, and by the fact that when one condor finds a carcass others swiftly fly in to join it. The explanation is relatively simple: if one condor, soaring high in the air, sees another suddenly drop to the ground, it knows that it has probably found food. As this condor glides over to join the first, its movements are seen by others, which converge on the same spot.

For some time scientists believed that condors located carrion by detecting the smell of decomposing matter. It is now accepted that most birds have a weak sense of smell. John Audubon, the American ornithologist, showed

CONE SHELL

THERE ARE MORE THAN 1,000 species of cone shell, marine snails named for the shape of their shells. As is the case with land snails, the shells consist of a tube wrapped round a central column. In cone shells the tube is flattened, making a long, tapering shell. The cone is formed by the large outside whorl of the tube, with its narrow, slitlike opening extending to the tip of the cone. The base of this cone is formed by the short, sometimes almost flat spire made by the exposed parts of the inner whorls. The surface of the shell is generally smooth and has a dotted or lace-like pattern of brown markings on a white ground, although the coloration varies according to species.

Cone shells are up to about 10 inches (25 cm) in length, the majority of species being much smaller. Attached to the foot of each cone shell is an elongated operculum, the horny lid or door that seals the aperture of the shell when its owner retires inside. When the snail is active, a pair of sensory tentacles and a long siphon protrude from the front of the body. Water is drawn in through the siphon by beating cilia and passed over the gills in the cavity within the shell.

Cone shells are found in tropical and sub-tropical waters, mainly in the western Atlantic, including the Caribbean and Gulf of Mexico, around the Philippines and Malay Archipelago, and across the Indian Ocean to East Africa. They also occur in the Red Sea and Mediterranean. Cone shells live in shallow water to a depth of several hundred feet. Some live on coral reefs, others in coral sand or rubble. They are active mainly by night, coming out to feed after lying up in crevices or under stones during the day, or burying themselves in the sand with only the siphon showing.

Cone shell eggs are laid in vase-shaped capsules of a hard, parchment-like material. The capsules are attached by their bases in groups to the coral or rock. They hatch in about 10 days.

Poison harpoons

Cone shells are carnivorous, feeding on worms, crustaceans, other mollusks and live fish such as blennies or gobies. Each species of cone shell has its own preferred prey, which it first paralyzes and then swallows whole. Capture of the prey is accomplished by means of a snoutlike proboscis armed with long, poisonous teeth. The teeth are

basically the same structures as those found on the radula, or tongue, of other snails, such as the abalone or banded snail of the genus *Haliotis*. In most snails the radula feature many small teeth, forming a file for scraping off particles of food, but in the cone shells there are only a few teeth, each up to ⅒ inch (1 cm) long. Each tooth is a long, barbed, hollow harpoon mounted on a mobile stalk. A poison gland is connected to the teeth by a long tube. The viscous, milky white poison is squeezed out by the contraction of muscles around the poison gland.

Cone shells detect scent particles secreted by their prey with a sense organ called the osphradium, which tastes the water as it is drawn through the siphon. A captive fish-eating cone shell will respond if water from a tank containing fish is put into its own tank.

A cone shell either tracks down its prey, if it is another slow-moving animal such as a worm or mollusk, or lies in ambush. When the victim comes within range, the cone shell uses its proboscis like a harpoon, to launch one or more of its teeth at the prey. This attack is followed by a secretion of poisonous chemicals, forming a cloud of nerve noxious gas that paralyzes the victim. If the prey is another mollusk, the paralysis causes it to lose its grip on its shell so the soft body can be drawn out.

Fish-eating cone shells, such as the elephant trunk cone, Conus striatus, *are sensitive to chemical trails left in the water by prey. When prey is nearby the cone shells extend their proboscis to locate the source of the chemical stimulus.*

Fish-eating cone shells bury themselves in the sand, extending the water siphon to smell the water for prey. When a fish comes close, the cone shell brings its proboscis out of its sheath and brings it rapidly down on to the prey. At the same time a single tooth is rotated outward and thrust into the victim's body. In a successful attack, the prey is paralyzed by the poison within seconds and the cone shell emerges from the sand, withdraws its proboscis and poison tooth and swallows the fish whole. The mouth at the end of the proboscis dilates and engulfs the victim's body, rapid muscular contractions forcing it down the gullet with the help of a lubricating secretion. Swallowing and digestion may take several hours, during which time the snail is not able to retreat into its shell. However, if the attack is unsuccessful, the cone shell abandons the prey, along with its tooth. The tooth usually breaks off when it has been used and another is brought forward for the next victim.

Dangerous mollusks

The poisonous teeth with which cone shells dispatch their prey present a great danger to anyone handling the living animals. Some cone shells may sting no more severely than a bee, but in other cases the pain may be excruciating and even fatal, death ensuing in 4–5 hours. One survey gives the death rate as 20 percent of all people stung, higher than that due to cobra or rattlesnake bites. The radula may extend several inches out of the shell and cone shells can sting so rapidly that the person being attacked may at first be unaware of the attack. Pain comes later, together with numbness, blurred vision and difficulty in breathing.

Cone shells inhabit shallow tropical and subtropical seas worldwide. The legate cone shell, Conus legatus, *is found in the South Pacific.*

CONE SHELLS

PHYLUM	**Mollusca**
CLASS	**Gastropoda**
ORDER	**Stenoglossa**
FAMILY	**Conidae**
GENUS	***Conus***
SPECIES	**More than 1,000, including glory of the sea, *C. gloriamaris*; and Mediterranean cone shell, *C. mediterraneus***

LENGTH
Up to 10 in. (25 cm); most species usually 2–4 in. (5–10 cm)

DISTINCTIVE FEATURES
Long, tapering shell with smooth surface and narrow, slitlike aperture; color variable

DIET
Mainly bottom-living animals, including mollusks, crustaceans, worms and fish

BREEDING
Age at first breeding: 1–2 years in most species; hatching period: about 10 days; larval period: about 7 days

LIFE SPAN
Probably 2–10 years

HABITAT
Coral reefs, coral rubble, rocks and sandy seabeds in shallow, coastal waters

DISTRIBUTION
Most species in tropical and subtropical seas; few species in Mediterranean

STATUS
Many species common

Cone shells

CONGER EEL

THERE ARE ALMOST 150 SPECIES of conger eels, which are stout-bodied, predatory marine fish. Probably the best-known species is *Conger conger*, of the North Atlantic and Mediterranean region. It is normally 5 feet (1.5 m) long, though it may reach 10 feet (3 m). Adults typically weigh about 100 pounds (45 kg), although there is an unconfirmed weight of 160 pounds (75 kg) for a specimen caught in 1940. Unlike freshwater eels, conger eels are scaleless. The body of *C. conger* is brown to dark slate in color and its underside may be golden or white. However, its coloration varies according to that of the seabed: on a sandy bottom the species is more or less colorless; on gravel or among rocks it becomes much darker.

The gill openings of conger eels are large and extend to the underside of the body. The mouth runs backward to below the level of the eye and is armed with rows of sharp teeth, one row in the upper jaw being set so close together that they form a cutting edge. The front pair of nostrils are tubular, each of the hind pair being opposite the center of the front edge of the eye. The eyes are large, and reminiscent of those of deep-sea fish. The pectoral fins are fairly large and the pelvic fins are lacking; the dorsal fin begins from above the pectoral fins and runs continuously along the back and into the anal fin.

Conger eels are often simply referred to as congers. The name conger derives from the Latin word *congrus*, meaning "sea eel."

Graceful predators

Conger eels are important marine predators, taking a wide variety of animal food. Major prey items include octopuses, large crustaceans and bottom-living fish. Crabs and lobsters are held in the strong jaws with a viselike grip and battered against rocks before being swallowed. Conger eels also feed on carrion.

A conger eel swims easily and gracefully. When cruising it is propelled largely by wavelike movements that pass down the dorsal and anal fins, from the front of the body to the rear. At greater speeds the body moves with a serpentine, side-to-side motion. A conger eel frequently turns on its side to swim using undulating movements of the body. When this action takes place at the surface of the water the eel's body appears as a series of humps. The fish spends the day lying among rocks or in crevices, sometimes on its back. It yawns periodically, though this action is probably connected with respiration rather than with fatigue.

Spawn once then die

In common with freshwater eels, conger eels spawn once and then perish. The conger eels of European Atlantic waters spawn in late summer near the Sargasso Sea at depths of 10,000 feet (3,000 m); *Conger conger* has another spawning ground in the Mediterranean. Before spawning conger eels stop feeding and become almost black in color, with very much enlarged eyes, especially in the male. The female lays up to 12 million eggs, each 2.5 millimeters in diameter. These eggs float in the intermediate layers of the ocean but occasionally reach the water's surface.

Conger eels (Conger verrouxi, above) often hide in crevices among seabed rocks, to wait for passing prey. Unlike freshwater eels, they lack scales.

Thin-headed larvae

The first eel larva was discovered in 1777 and belonged to a freshwater eel. The Italian naturalist Giovanni Antionio Scopoli, who discovered the larva, named it *Leptocephalus* ("thin-head") and believed that it was a new kind of fish. The first conger eel larva was found in 1763 by the British naturalist William Morris, but was not described until 1788, when the German scientist Johann Gmelin named it *Leptocephalus morrisii*. It was not until 1864 that this was recognized to be the larval stage of a conger eel. Even then, the distinguished ichthyologist Albert Gunther argued that the larva was an aberration, a species that had been arrested in its development. In 1886, however, the French zoologist Yves Delage proved beyond doubt that *Leptocephalus morrisii* was a larval conger eel by watching it change into a young conger in an aquarium. Later in the 19th century the true nature of what are now called the leptocephali became more firmly established when the Italian naturalist Raffaele managed to keep the eggs and larvae of five species of eels alive in aquaria.

The leptocephali of *Conger conger* lose their larval teeth on reaching coastal waters, and change into young eels. Their bodies becomes rounded and eel-like and their length drops from 5 to 3 inches (13 to 7.5 cm). Until the young conger eels reach a length of 15 inches (38 cm) they are a pale pink. After this they gradually take on the dark adult color. When about 2 feet (60 cm) long they move down the continental shelf into water up to 650 feet (200 m) deep, and spend most of their time on the seabed.

Conger eels hunt mainly by night. When hunting some species develop a pattern of light and dark bands that makes it harder for prey to see them.

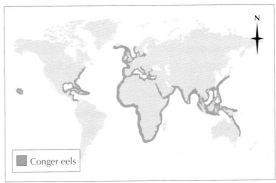

Conger eels

COOT

COOTS SWIM WELL, using the large lobed flaps on each of their toes as paddles. These leave the coots' toes free, unlike the webbed feet of ducks, which means that coots are also very nimble on land. Another feature particular to coots is the hard shield on the forehead, which is an extension of the bill. In the horned coot, *Fulica cornuta*, of South America, the shield is replaced by wattles and a fleshy horn, or caruncle.

The 11 species of coot are members of the rail family, Rallidae, which includes the rails, crakes, gallinules and moorhens. The greatest diversity of coots is found in South and Central America, where seven species occur. These include the horned coot and the giant coot, *F. gigantea*, which live on alpine lakes high in the Andes Mountains. Three species of coot are very similar in appearance, with blackish plumage and brilliant white bills; these are the American coot, *F. americana*, which ranges from central Canada to Colombia; the European, or common, coot, *F. atra*, which is found from the British Isles east to Japan; and the crested, or knob-billed, coot, *F. cristata*, which ranges from southern Spain to southern Africa. The endangered Hawaiian coot, *F. alai*, is closely related to the American coot, from which it was recently separated by ornithologists.

Diving for food

Coots live in a wide variety of freshwater habitats, from small ponds and marshes to fairly large bodies of water. Outside the breeding season coots gather in flocks of hundreds or thousands and may visit coastal waters.

During the breeding season coots keep to beds of reeds and other aquatic plants where they can frequently be observed threading their way through the stems. Sometimes they come out of the cover and can be seen crossing the water in a straight line, headed for another clump of reeds. When alarmed, coots run across the water with wings beating, leaving a trail of splashes, to subside with a crash and then disappear into vegetation. At other times they dive, upending and disappearing below the surface leaving scarcely a ripple. The European coot has been recorded staying underwater for almost 30 seconds.

Coots eat mainly water plants, diving down and bringing up masses of weed that are eaten at the surface. They also forage on land, eating grasses, plants and even acorns. This diet is supplemented by animal food, including fish, tadpoles, aquatic insects, mollusks and worms. Occasionally coots raid the nests of gulls and grebes, piercing the eggs to sip the contents.

Coots are not noted for the power of their flight. They are rarely seen in the air and take some time to build up speed on the water. The giant coot has lost the power of flight altogether and must travel between mountain lakes on foot. However, once airborne the remaining 10 species of coot are capable of sustained flight, and several species are migratory.

Aggressive birds

Quarrelling is a feature of the European coot's life that has often been remarked on. Both in the winter flocks and during the breeding season when pairs defend their territories, coots can be seen fighting, sending up sheets of water but rarely coming to blows.

The first sign of antagonistic behavior is the adoption of an aggressive posture by one or both coots, with head down and wings arched. If one does not retreat, the two coots swim together and abruptly erupt into dramatic bouts of water-throwing, sitting back on their tails and then splashing water over one another by beating with their wings and feet. The display is usually over in a few seconds, after which one of the coots may retreat, pursued by the other.

Most species of coot have a horny extension of the bill on the forehead. This feature is known as a shield and is red brown in the American coot (above).

Coots have fleshy lobes on each toe that help them paddle in water and walk over soft ground. Pictured is the European coot.

AMERICAN COOT

CLASS	**Aves**
ORDER	**Gruiformes**
FAMILY	**Rallidae**
GENUS AND SPECIES	***Fulica americana***

LENGTH
Head to tail: 12–14½ in. (30–37 cm); wingspan: 24–28 in. (60–70 cm)

DISTINCTIVE FEATURES
Compact, stocky body; large feet with fleshy lobes; mainly black plumage; pure white bill; reddish shield on forehead; red eyes

DIET
Mainly water weed, grasses and other plants; also invertebrates, small fish and bird eggs

BREEDING
Age at first breeding: 1 year; breeding season: April–August; number of eggs: usually 8 to 12; incubation period: 23–24 days; fledging period: 49–56 days; breeding interval: up to 3 broods per year

LIFE SPAN
Up to 9 years

HABITAT
Freshwater marshes, lakes and ponds; also on coastal waters in winter

DISTRIBUTION
Central Canada south to Colombia and western Venezuela; Caribbean islands

STATUS
Common in much of range

American coot

Floating island nests

Typically, coots make their nests among tall water plants on the margins of lakes and ponds, building up a mass of dead reeds, stems and leaves often rising 1 foot (30 cm) above the water level and having a slipway for easier access. If the water level rises, more vegetation is added to keep the eggs dry. The giant coot makes a floating platform of vegetation, sometimes anchored to the stems of living water plants; the horned coot, which often nests nearby, often sites its nest on natural hummocks, using stones as a base.

Caring for the chicks

The American and European coots breed in April–August. Usually 8 to 12 eggs are laid, but there may be as many as 20. Both parents incubate the eggs, which hatch in just over 3 weeks. At first the female broods them while the male brings food, but after a few days the chicks leave the nest, returning at night to be brooded. A number of predators attempt to steal coot eggs; in Britain gulls and crows in particular are important egg predators.

Coots defend their nests by sitting on them, but the eggs are at risk whenever they are exposed. At first the parents do not recognize their own chicks independently and will accept and feed any chick about the same size as their own. Chicks considerably larger or smaller than their own, however, are attacked and even killed if they do not retreat. Later, when their own chicks are approximately 2 weeks old, the parents recognize them as individuals and no other chick is allowed in the territory.

The young chicks are clad in a sooty down but their heads are brilliantly colored. Around the bill, which is white with a black tip and shading to vermilion on the frontal shield, is a patch of red down. The sides of the face are orange, the crown is blue and the nape is red, orange or black. When they are 1 month old the chicks begin diving for their own food and the down changes to a uniform blackish color, relieved by dirty white underparts.

COPPER BUTTERFLY

THE WINGS OF THE COPPER butterflies have the color and luster of polished copper and are marked with dark spots and bands, occasionally with blue or purple patterns as well. The coppers are a group of small butterflies in the family Lycaenidae, and are closely related to the blue and hairstreak butterflies.

Copper butterflies are widely distributed in the temperate and cold regions of the Northern Hemisphere, from North America east to Greenland, Europe, Asia and Japan. There are also three species of copper butterfly in temperate New Zealand. From their close genetic resemblance, scientists presume that the ancestors of these species arose from the same stock as those in the Northern Hemisphere and became separated over time. Copper butterflies favor open spaces and are often found close to urban areas, such as beside roads and on vacant lots.

The caterpillars (larvae) are shaped like wood lice and, in the majority of species, feed on various kinds of dock and sorrel. Like the larvae of many of the blue butterflies the larvae of some coppers are attended by ants, which are attracted by a sweet, nutrient-rich secretion that the larvae produce. In the case of copper butterflies this exudes all over the body of the larvae, unlike the larvae of many other butterflies in the family

Lycaenidae, which have a single orifice connected with a special gland. The ant-lycaenid association is symbiotic (of mutual benefit). Many copper butterfly larvae mimic ant odors, thereby largely avoiding predation by ants; in some cases the larvae actually enter ant nests and are protected by the colony. Ants benefit from the sweet food rewards provided by the larvae.

The small copper

Many species of copper butterfly are abundant and widespread. The small copper, *L. phlaeas*, has an enormous range incorporating a large part of North America and extending from Europe right across Asia to Japan. The small copper is divided into distinct subspecies in different parts of its range but all are of similar appearance. One subspecies ranges farther north than almost any other butterfly, onto Ellesmere Island; it is one of the five species of butterfly in Greenland.

The small copper is an active and occasionally aggressive little butterfly. Males and females are very much alike in appearance, a rare characteristic in the family Lycaenidae. Both sexes have golden-red forewings edged with brown and bearing black markings; the hind wings are brown, banded with red and feature two short tails. Male small coppers establish territories and

Due to its iridescent copper-orange coloration, the female large copper butterfly is often confused with both sexes of the small copper butterfly.

try to chase all other butterflies away, flying out and attacking individuals of their own species as well as those of other, larger species, though they are incapable of injuring one another.

The life cycle of the small copper can be so short that there can be three generations of adults in a single summer, though the time taken to complete the life cycle varies with different environments. The larva of the small copper feeds mainly on curled dock and sheep sorrel. It is green with a brown line along the back and is covered with short grayish hairs. It is not attended by ants. The pupa is pale brown or greenish and attached to a leaf or stem of the food plant. The small copper overwinters as a larva, and the adult butterfly is on the wing from April to October, depending on the latitude.

The large copper

Another widespread species of copper butterfly is the large copper, *Lycaena dispar*, which is found from western Europe east to easternmost Siberia. The male and female look very different. In the male the upper side of all four wings is brilliant red-orange with only narrow dark borders and small central dots. The female has dark markings not unlike those of the small copper.

The large copper was discovered in Britain a little before 1800 in the marshes of East Anglia in the southeast of the country. This habitat was rapidly shrinking due to artificial drainage. Unrestrained collecting led to the demise of the British subspecies, the last specimens being taken in 1847 or 1848. Though this race is now extinct the large copper is still found in many parts of

Europe and Asia. The great water dock is the food plant of the large copper. The larva is green and slightly flattened in appearance. It hibernates when young, feeds in the following spring, and the adult butterflies appear in July–August.

Two other species of copper butterflies, the scarce copper, *L. virgaueae*, and the purple-edged copper, *L. hippothoe*, are both palaearctic in distribution, ranging over an area including Europe, Asia north of the Himalayas, northern Arabia and Africa north of the Sahara.

The small copper has a vast range that includes most of Europe, Asia, Japan and North America.

SMALL COPPER

PHYLUM	**Arthropoda**
CLASS	**Insecta**
ORDER	**Lepidoptera**
FAMILY	**Lycaenidae**
GENUS AND SPECIES	***Lycaena phlaeas***

LENGTH
Adult wingspan: 1–1½ in. (25–35 mm)

DISTINCTIVE FEATURES
Adult: forewings bright copper orange with several dark brown spots. Larva: green body, sometimes with pinkish spots.

DIET
Adult: flower nectar. Larva: plants of family Polygonaceae, such as sorrel and dock.

BREEDING
Breeding season: summer; larval period: 120–150 days; breeding interval: up to 3 generations per year, depending on latitude

HABITAT
Wide variety of grasslands, including tundra, meadows, heaths, gardens and waste ground

DISTRIBUTION
Much of North America, south to Kentucky and Virginia; Greenland; Europe; Asia; Japan

STATUS
Common

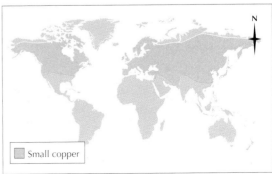

Small copper

CORAL

ORALS ARE AN EXTREMELY diverse group of marine invertebrates in the phylum Cnidaria. They are similar to anemones in many respects, although they are supported on a hard, chalky skeleton, which has a stonelike, horny or leathery consistency. This skeleton is covered with a continuous layer of flesh, from which spring cylindrical structures of tissue called polyps.

Apart from corals, the phylum Cnidaria also includes anemones, jellyfish, sea fans, sea pens and the common laboratory *Hydra*, and its classification is highly complex. There are a number of orders of corals, the largest of which is the stony, or true, corals (Scleractinia). Other important coral orders include the black and thorny corals (Antipatharia), the horny corals (Gorgonacea), the blue corals (Helioporacea), the fire corals (Milleporina) and the soft corals (Alcyonacea).

Structure

The stony corals may be either solitary or colonial. Solitary species are distinguished by a single polyp living on its own, seated in a chalky cup or on a mushroom-shaped chalky skeleton. Colonial species comprise a sheet of tissue, formed by hundreds or thousands of polyps, which covers the skeleton. Coral polyps have an extremely diverse structure; common arrangements include tree, cup and dome forms. The overall structure of entire coral colonies, which are created by many individual polyps, is equally diverse. Some colonies are made up of flattened plates; others consist of branchlike structures resembling a stag's horns.

Coral polyps are attached to the coral skeleton at one end and have a mouth at the tip of the opposite, free, end. The mouth is surrounded by tentacles armed with stinging cells, which paralyze prey. The number of tentacles per polyp is six or a multiple of six in the stony corals, according to species, and eight in soft corals.

Soft corals are usually quite flexible, though they are treelike in form and the centers of the stems and branches are strengthened by a red or black chalky material. This material, stripped of its flesh, is commercially valuable. Related to these precious corals are the sea fans, the stems and branches of which are strengthened by a flexible horny material. The organ-pipe coral, of the order Stolonifera, consists of a mass of vertical tubes joined at intervals along their length by thin horizontal plates. Its skeleton is reddish purple and the polyps are a pale lilac color. When expanded these take on the appearance of delicate flowers.

Corals (above) feed in the same way as sea anemones. Their extendable tentacles contain stinging cells that paralyze tiny prey and drag it down into the waiting mouths.

Tropical reef builders

Large accumulations of corals of various species on the seabed are known as reefs, and coral reefs are frequently compared to tropical rain forests on account of the quantity and diversity of other living organisms they support. Hundreds of species of fish, as well as a wide variety of algae, mollusks, crustaceans and other invertebrates, may thrive on a single reef. A smaller number of marine mammals and reptiles, including green, loggerhead and hawksbill turtles and certain sea snakes, also occur on coral reefs.

Corals are found in many of the world's seas, though most species are found in the Tropics; reef-building corals in particular favor warm waters. Thousands of miles of tropical shores, especially in the Indian Ocean, are edged with reefs. In places barrier reefs are formed offshore, and in midocean, especially in the Pacific, there are ring-shaped atolls made of living coral, topping accumulations of dead coral skeletons. Atolls may extend downward into the ocean to depths of up to 1 mile (1.6 km).

Stony corals occur in a wide variety of forms, including fanlike structures and brainlike whorls. The form of each coral depends on the growth pattern of the colony and on the arrangement of the individual polyps within it.

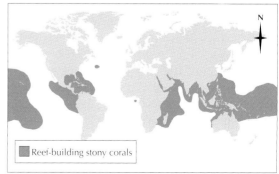

STONY CORALS

PHYLUM	**Cnidaria**
CLASS	**Hexacorallia**
ORDER	**Scleractinia**
GENUS	**Many, including *Fungia, Porites, Heteropsammia, Diploria* and *Acropora***
SPECIES	**More than 2,500**

LENGTH
Solitary species: about ½ in. (1 cm); colonial species: up to 6½–10 ft. (2–3 m)

DISTINCTIVE FEATURES
Stonelike skeleton covered with polyps (cylindrical structures of tissue); each polyp has tentacles in multiples of 6; structure and coloration of polyps extremely diverse, with many variations on basic tree, cup and dome forms; overall shape of colonial species (comprising many polyps) also highly diverse

DIET
Filter plankton from water; also feed on zooxanthellae (single-celled algae) growing inside their own tissues

BREEDING
Breeding season: all year or seasonal, according to species; larval period: several days to several weeks

LIFE SPAN
Probably 10–100 years

HABITAT
Most species in warm, shallow waters; a few species in cool waters and at depths of up to 19,500 ft. (6,000 m)

DISTRIBUTION
Widespread in tropical seas; some species in temperate seas

STATUS
Many species common, but increasing numbers of coral reefs dead or damaged

Reef-building stony corals

Birth of a reef

Reef-building corals are found north and south of the Equator, about as far as the 25th line of latitude, a zone in which the temperature of the sea never falls far below 65° F (18° C). Once a year, the corals engage in sexual reproduction. During a short period of hours or days during the summer, when the water temperature, light intensity, tidal range and cycle of the moon are favorable, all the breeding corals within a particular area release thousands of eggs, each polyp producing one egg bundle. The eggs differ in color according to the color of the parent. Some corals produce both male and female sex cells; others release only cells of one sex.

When the eggs are fertilized they become larvae called planulae. These join the plankton in the upper levels of the sea for a short period, and may be carried by ocean currents to a new site some distance away from their parent. The larvae then descend to the seabed, where they grow into coral polyps.

A small lump appears on the side of each polyp. This is a bud; as it becomes larger, a mouth appears at its free end and a crown of tentacles grows around the mouth. The bud grows until it is the same size and shape as the parent, but remains connected to it. By repeated budding of the parent stock, and of the new growths formed from it, a colony numbering up to many thousands of polyps is formed.

Between them the coral polyps build a common skeleton, which may eventually grow to a height and width of several feet. The tissues of each polyp's body walls spread out over the edges of the limestone skeleton and join with those of the adjacent polyp. The polyps of a single coral colony are clones of one another and are all connected via a fold of the stomach wall called the gastrodermis; a colony of coral polyps therefore effectively has a communal stomach. Sometimes, branched coral species reproduce by fragmentation: broken pieces of coral may become reattached to the reef and subsequently continue growing to form new chains.

Living animal traps

All corals, whether solitary, colonial, reef-building, stony or soft, feed like sea anemones. Their tentacles paralyze small swimming animals and then push them into the mouth at the center. In reef-building corals the polyps are withdrawn during the day, so that the surface of each coral mass is fairly smooth. As night falls and planktonic animals rise into the surface waters, the polyps and their tentacles become swollen with water; this process is controlled by cilia (hairlike projections) on the skin, which beat rhythmically to create a current of water into the mouths. The

polyps now stand out on the surface of the coral, their tentacles forming a semitransparent pile containing many mouths, which are waiting to receive prey. The seemingly inert coral has been converted into a huge trap for passing small animals, underlining the coral's close relationship with anemones. A secondary reason for nocturnal activity in corals is the avoidance of predators. Most polyp-feeding fish, such as parrot fish, are active by day.

Mutual relationship with algae

Corals also rely heavily on zooxanthellae for much of their nutrition. Zooxanthellae are microscopic, single-celled algae and occur in huge numbers in the tissues of corals. They take up carbon dioxide and other waste products from their hosts and, being plants, are capable of photosynthesizing. In return much of the product of this photosynthesis is used by the corals for nutrition; the zooxanthellae also produce oxygen, which likewise can be used by the corals. It is thought that many corals obtain a greater proportion of their food requirements from zooxanthellae than from filter feeding.

The dependence of zooxanthellae on light for photosynthesis explains why corals are mainly limited to very shallow waters. The variety of corals present in the top 33–50 feet (10–15 m) of water greatly exceeds the range of species that occurs below this level, as the light intensity is so much greater near to the water surface. Reef-building corals are absent from river deltas and from the coasts of West Africa and eastern South America, because these waters are clouded by silt from the Congo and Amazon rivers.

When viewed from above, the mouths of individual coral polyps are clearly visible, each surrounded by a ring of tentacles.

CORAL REEF

Almost 5,000 different types of fish, more than 20 percent of all the known species, are found on coral reefs.

CORAL REEFS CONTAIN THE most colorful and biologically diverse animal communities known to science. Only one of the the world's 33 animal phyla is not represented in this biome. It is estimated that nearly half a million species of animal occur on reefs, and new species are discovered there every year.

The extraordinary variety of fish and marine invertebrates living on coral reefs interact in a complex web of competition, predation and symbiosis (relationships of mutual benefit). One of the main factors that has enabled such a wide range of organisms to evolve on coral reefs is the structure of the reefs themselves. A typical reef consists of a network of plateaus, cliffs, caves and canyons, the surfaces of which feature an array of ledges, crevices and holes, of all shapes and sizes. The reef superstructure creates a host of microhabitats for plants and animals.

Need for stable conditions

Coral reefs are found only in seas where the water temperature is at least 61° F (16° C) and form most successfully in waters of 65–86 °F (18–30 °F). Corals do not respond well to seasonal temperature variations, hence their concentration in tropical seas, where the conditions are relatively constant all year. They also require perfectly clear, strongly saline water in which to grow. This means that coral reefs are absent from coastal areas where large rivers dilute the sea with fresh water, thereby clouding the sea with sediments eroded from the river basins. It is thought that coral reefs cover about 230,000 square miles (600,000 sq km) of the earth. Most are located in the tropical Indian Ocean, eastern Pacific and the Caribbean.

Types of reef

The coral reefs around the world can be divided into four major categories: fringing reefs, barrier reefs, coral atolls and patch reefs. Fringing reefs are the simplest and most common form. They are underwater platforms of coral that vary in thickness; they run along the coastline and slowly grow out from the coast. Barrier reefs are similar in appearance, but the lagoon area enclosed between the shore and the reef is deeper than that bordered by fringing reefs. Most barrier reefs are linear and run parallel to the shoreline;

CORAL REEF

GROWING CONDITIONS
Warm, very clear waters of consistent high salinity, usually at depth of less than 50 ft. (15 m). Water temperature: 65–86° F (18–30° C).

STRUCTURE
Colonies of corals, coral-like animals, algae and sponges provide reef foundations

LOCATION
Tropical seas between 25° N and 25° S; absent from much of Atlantic and parts of eastern Pacific; generally (except coral atolls) found near to coastlines

STATUS
Many reefs damaged to varying degree, especially in Philippines, Malaysia, Indonesia and parts of Caribbean

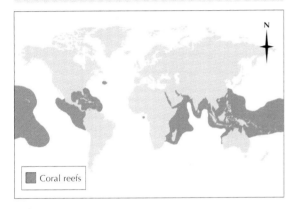

Coral reefs

occasional breaks in the coral connect the sheltered lagoon with the open ocean. Others are nearly circular and may surround an island. The Great Barrier Reef off Queensland in northeastern Australia is over 1,250 miles (2,000 km) long and covers an area of more than 80,000 square miles (200,000 sq km). The world's second largest barrier reef skirts most of Belize's coastline, which is 175 miles (280 km) long.

Coral atolls may occur many miles from land in deep water. They are ring-shaped reef islands that have grown above the water's surface, enclosing a shallow lagoon, which is usually only 65–200 feet (20–60 m) deep. At the center of each coral atoll lies a submerged volcanic island. There are several hundred atolls in the Maldive Islands, in the Indian Ocean. Three of the four largest Caribbean atolls are in Belizean waters.

Patch reefs form when isolated colonies of coral grow equally in all directions. This growth pattern creates small, irregularly shaped reef islands, which often occur in the lagoons of barrier reefs or coral atolls.

Darwin's theories of reef-formation

In 1831 the English scientist Charles Darwin sailed with H.M.S. *Beagle* to investigate islands in the Pacific and Indian Oceans. On his return to England in 1836, he proposed a new theory of reef growth based on the existence of underwater volcanoes. Darwin outlined his theory in *The Structure and Distribution of Coral Reefs*, published in 1842. He argued that millions of years ago the volcanoes emerged above the surface of tropical

A single colony of living coral may be used by hundreds of other species in one way or another, some of which may not occur anywhere else.

Atolls are created when a coral reef adopts a ring shape, surrounding a lagoon (bottom right of photograph). Part of the reef may later be colonized by plants, effectively becoming an island (above, center).

oceans to become islands. When volcanic activity ceases, an island begins to erode, becoming fractured with jagged ridges and peaks, and a fringing reef grows around its shoreline. Erosion and movements on the seabed cause the island gradually to sink again; at the same time, the corals in the fringing reef continue to grow upward. Thus, even when the highest peak on the island has sunk, a circular coral atoll, or reef, remains just below or at the water's surface. When the island sinks sufficiently to create a lagoon between the land and the reef, a barrier reef is created.

Contemporary scientists criticized Darwin's theory, especially his belief that entire islands could rise or sink. However, his hypothesis has subsequently been verified through the study of the movements of the earth's crust, known as plate tectonics, which geologists use to explain the creation and destruction of islands.

Reef communities

The organisms present in the coral reef biome can be divided into three broad groups, each of which contains thousands of species. The reef itself consists of sedentary colonies of hard and soft corals, coral-like animals, sponges and algae. There are more than 350 species of reef-building corals in the Great Barrier Reef alone. The second group of species found in the biome is called cryptofauna and consists of animals that either bore into the reef structure or inhabit its surface. Included in the cryptofauna are sponges, flatworms, polychaete worms, sea anemones, brittle

stars, starfish, sea cucumbers, sea urchins, bivalve mollusks, shrimps, crabs, lobsters and octopuses. A single colony of dead coral was once found to contain 220 species of boring cryptofauna alone, which illustrates the tremendous diversity of this group of reef animals. The third group of species found in the coral reef biome consists of fish, squid, cuttlefish, jellyfish, turtles and a range of other animals.

Diversity of reef fish

Fish are among the most important residents of coral reefs, acting as both predators and prey. They live in every part of a reef and have evolved many different lifestyles, specializing in particular food types. Many herbivorous species, such as damselfish, surgeonfish and rabbitfish, aggressively defend small territories on the reef in order to have exclusive access to the rich patches of algae on which they feed. Parrot fish and pufferfish also graze on algae, but have strong jaws and chisel-like teeth that enable them to nibble corals to extract the single-celled algae found inside the tissues of the coral itself. A large number of schooling fish species feed on plankton in the shallows over coral reefs. The diurnal (day-active) plankton-feeders include angelfish, groupers and triggerfish; nocturnal species include cardinal fish, squirrelfish and soldierfish. Invertebrate predators are probably the largest single group of reef fish, and include eagle rays, which take mollusks and crustaceans. Groupers, barracudas, eels, snappers and sharks are among the many reef fish that prey on other fish.

Species of fish that have broadly similar diets can avoid competition for food by keeping to different levels of a coral reef. On reefs in the Maldive Islands, for example, the Picasso triggerfish inhabits sheltered lagoons, while the titan triggerfish is found at the seaward edges of reefs and the clown triggerfish lives at the lower levels of the outer reef slopes. Tagging experiments have shown that some species of fish remain in the same location for long periods of time.

Ancient relationships

Coral reef environments can be very old, dating back 50 million years or more. As a result there has been time for complex interactions to evolve between various species of reef fish and between reef fish and invertebrates.

Coral reefs are especially rich in examples of symbiotic relationships. For instance, there are a number of brightly colored "cleaner fish" that eat the parasites, loose scales, mucus and scraps of food lodged on larger fish. Wrasses (Labridae) are the dominant cleaner fish on reefs in the Indian and Pacific Oceans; neon gobies (*Gobiosoma*) are the most familiar species in the Caribbean. Cleaner fish establish cleaning stations at regular intervals along the reefs, around which other fish congregate in order to be tended.

Clownfish are a well-studied example of fish that have evolved symbiotic relationships with invertebrates, in this case sea anemones. The clownfish gain protection from predators by living among the anemones' stinging tentacles, yet escape injury; scientists believe that chemical secretions contained within the mucus in the external coat of the clownfish inhibit the stinging action of the tentacles. In return, the clownfish remove debris from among the tentacles, and may even lure other fish, potential prey, within reach of the waiting anemone.

A fragile environment

For millions of years coral reefs have been subjected to tremendous variations in climate, and they have always experienced local advances and retreats. Hurricanes are a major naturally occurring threat to reefs, but in time the hurricane damage is repaired by coral regrowth. However, human activity is placing ever greater pressure on this fragile biome.

The main threats currently facing coral reefs are overfishing, unregulated tourism, water pollution, mining and global warming. Overfishing disrupts the delicate balance of coral reef food webs. Fishing with cyanide and dynamite has become common in several regions, particularly in Southeast Asia, and is very destructive to the reefs themselves. The growing popularity of scubadiving has led to the deterioration of reefs, especially in the Red Sea; damage is caused by boat anchors and divers' flippers. In some areas coral is mined to provide building materials and to enlarge harbors. Pollution, in the form of excessive silt, sewage, oil and industrial chemicals, reduces water clarity and the rate at which corals grow. The most serious threat to coral reefs, however, is probably the rise in sea levels and temperatures associated with the phenomenon of global warming, which causes the bleaching and eventual death of the corals.

The conservation of coral reefs is of paramount importance. Although this biome covers less than 2 percent of the ocean floor, it is home to about 25 percent of all marine species.

CORAL SNAKE

The Arizona coral snake is less than half the size of the almost identical common coral snake, and it is seen much less frequently.

CORAL SNAKE IS THE NAME given to many strikingly colored snakes with bold patterns of rings running round the body and tail. The body is slender, and there is no pronounced distinction between head and neck. In North and South America there are several genera of true coral snakes, which are close relatives of the cobras, as are the Oriental coral snakes belonging to the genus *Maticora*. In South Africa some members of the genus *Aspidelaps* are called coral snakes; they are similar in appearance and habits to their American relatives.

Of the 60 species of coral snake found in the Americas, only two extend as far north as the United States. The two North American species have prominent rings around the body in the same sequence of black, yellow or white, and red. The Arizona, or Sonoran, coral snake, *Micruroides euryxanthus*, is small, having a maximum recorded length of about 20 inches (50 cm). The larger, common coral snake, *Micrurus fulvius*, occasionally reaches 4 feet (1.2 m). The range of the common coral snake extends north from northeastern Mexico, through eastern Texas to the low-lying country of Kentucky and North Carolina and south to Florida. The Arizona coral snake lives in the arid lands of Arizona, New Mexico and northwestern Mexico. Other American species of coral snake range as far south as central Argentina.

Brightly colored banding is not invariable in coral snakes. Those in the genus *Leptomicrurus* have long, thin bodies and short tails, which are dark on the upper side and which have yellow spots underneath.

Feeding and breeding

Coral snakes are nocturnal, lying up during the day in runs under stones and bark or in mossy clumps, though they are sometimes active by day if it has been raining. The jaws of coral snakes do not open very wide and the species can therefore eat only slender prey. Important prey items include small lizards, other snakes and frogs. On Trinidad, mongooses introduced to control the numbers of snakes have not adversely affected the island's population of coral snakes.

The female common coral snake lays 3 to 14 soft, elongated eggs in May or June, in a hollow in the earth or under a log. When the young snakes hatch, after 10–12 weeks, they measure 7–8 inches (18–20 cm) and have pale skins, the colors of which become more intense with age.

Poisonous but rarely dangerous

A coral snake approaches its prey slowly, sliding its head over the victim's skin. The fangs are short and to inject a lethal quantity of venom the snake chews the flesh, lacerating the skin to force in a large amount of poison, which acts on the nervous system and has a powerful effect. In Mexico the common coral snake is called the "20-minute snake" as its bite was traditionally thought to be fatal to humans within that period, though 24 hours is probably a more accurate figure. Relatively few human deaths have been attributed to coral snake bites. If a coral snake senses danger, its initial impulse is to coil itself up and raise its tail, which it moves in an attempt to distract a would-be predator's attention from its more vulnerable head.

Warning colors?

Parallels can be drawn between the brightly colored markings of coral snakes and the bright stripes typical of certain insects, especially some bees and wasps. Conspicuous coloration is a feature of many animals that are toxic or have defensive venom, although many predators of snakes, including mammals, are color-blind. The dramatic light and dark bands of coral snakes would be noticeable to even color-blind species and may serve as a visual warning that these snakes are venomous.

It has been suggested that the banding on coral snakes may serve as a disruptive pattern, breaking up the outline of a coral snake's body, thereby making it less readily seen by would-be predators. Another theory is that the bright coloration may act as sexual signaling. Intense,

NORTH AMERICAN CORAL SNAKES

CLASS **Reptilia**

ORDER **Squamata**

FAMILY **Elapidae**

GENUS AND SPECIES **Common coral snake,**
Micrurus fulvius; **Arizona coral snake,**
Micruroides euryxanthus

ALTERNATIVE NAMES
Micrurus fulvius: eastern coral snake;
Texas coral snake. *Micruroides euryxanthus*:
western coral snake; Sonoran coral snake.

LENGTH
Micrurus fulvius: usually up to 4 ft (1.2 m).
Micruroides euryxanthus: up to 20 in.
(50 cm).

DISTINCTIVE FEATURES
Slender body; small, inconspicuous head;
striking pattern of black, white, yellow
and red bands

DIET
Mainly other snakes, lizards and frogs;
probably also some insects

BREEDING
Micrurus fulvius. Breeding season: usually
May–June; number of eggs: 3 to 14;
hatching period: 70–84 days.

LIFE SPAN
Not known

HABITAT
Micrurus fulvius: open woodland, scrub
and deserts. *Micruroides euryxanthus*:
semiarid areas.

DISTRIBUTION
Micrurus fulvius: northeastern Mexico and
eastern Texas east to North Carolina and
Florida. *Micruroides euryxanthus*: Arizona,
New Mexico and northwestern Mexico.

STATUS
Generally locally common

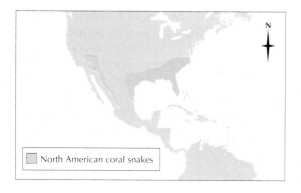

North American coral snakes

gaudy colors may be physiologically costly for a snake to maintain and, if so, it follows that a colorful individual is likely to be fit and healthy. Female coral snakes may therefore use body coloration as a guide to the quality of a potential mate. Parasites and disease may weaken body coloration; pale, faded colors may be a signal that an individual is not genetically strong.

Snake mimics

In the insect world, some harmless insects, such as certain flies and hoverflies, mimic the color patterns of harmful species. By doing so, they gain protection because predators learn to associate the color with an unpleasant taste and refrain from taking insects of that color. Coral snakes may also have their mimics, for there are nonpoisonous snakes with brightly colored rings in the Americas, Africa and Asia.

In the United States some of the reports of coral snakes found in unusual places may be explained as erroneous sightings of other, non-venomous species. These include the common king snake (*Lampropeltis getulus*), the shovel-nose snake (genus *Chinactis*) and several species of milk snakes. All of these snake species, however, have a different sequence of colored bands to that of the common and Arizona coral snakes. In the North American coral snakes each red band always has a yellow or white band on either side. In the "mimics" each red band has a black band on either side.

Scientists disagree as to the purpose of the colored bands typical of coral snakes. They may deter would-be predators, provide camouflage or play a part in mate selection. Pictured is Micrurus nigrocinctus of Costa Rica.

CORMORANT

Cormorants have long necks and powerful, daggerlike bills for spearing fish. This is the great cormorant, which occurs almost worldwide.

THE NAMES CORMORANT AND SHAG are applied more or less indiscriminately to the 42 species of the family Phalacrocoracidae. Cormorants are pelicanlike waterbirds with fully webbed feet and a throat pouch that can expand to accommodate and position large fish for swallowing. The pouch also helps cormorants to prevent themselves overheating. Primarily dark birds, cormorants soon warm up in direct sunlight. By panting and rapidly fluttering the throat pouch, cormorants cool the blood passing through the network of capillaries in the pouch, which helps the birds to cool down. Another characteristic of cormorants is a distinctive feather structure that enables water to penetrate easily and expel air trapped among the feathers, which is vital for waterbirds that hunt by diving: high buoyancy due to trapped air would hinder underwater swimming.

Cosmopolitan waterbirds

Cormorants are represented in coastal waters all over the world, except around the islands of the central Pacific. They also frequent suitable inland waters, including rivers, lakes, reservoirs and swamps, but do not occur in northern Asia.

The most common and widespread species of cormorant in North America is the double-crested cormorant, *Phalacrocorax auritus*, which is glossy black with naked orange-yellow skin on the face and blue eyes. Early in the breeding season double-crested cormorants from the West Coast have white tufts on the head and white plumes on the neck; eastern birds have darker, much less conspicuous tufts and plumes. The common cormorant, *P. carbo*, is the largest of all the cormorants; it is also known as the great cormorant in North America and as the black cormorant in Australia and New Zealand. The number of great cormorants found during the summer in the maritime provinces of eastern Canada is increasing. In winter the species is found along the eastern seaboard of the United States as far south as South Carolina.

Four other species of cormorant are found in North America. The neotropic, or olivaceous, cormorant, *P. olivaceus*, is limited to the Gulf Coast of Texas. Brandt's cormorant, *P. penicillatus*, and the pelagic cormorant, *P. pelagicus*, both occur on the Pacific coast. The red-faced cormorant, *P. urile*, is confined to southern Alaska.

Rarely far from shore

Many cormorants are closely associated with the sea but, unlike seaducks, auks and penguins, are rarely found far from shore. Most species live near shallow coastal waters, or inland on lakes and rivers. The most cosmopolitan species is the great cormorant, which is found in much of Europe, Asia, Africa, Australasia and eastern North America. The species with by far the most restricted range is the flightless cormorant, *Nannopterum harrisi*, which breeds only on two islands of the Galapagos Archipelago, to the west of Ecuador in the eastern Pacific. A large cormorant, it cannot fly and resembles penguins in its flipperlike wings and dense plumage. It is one of eight cormorants classed as vulnerable by the I.U.C.N. (World Conservation Union), all of which live in the Southern Hemisphere.

Cormorants are generally strong fliers and soar in air currents, but usually fly low over the surface of the water. All cormorants swim well, floating low in the water, sometimes with only head and neck showing. To submerge, they either jump up and plunge in headfirst, or else sink beneath the surface. The longest recorded cormorant dive lasted over 70 seconds and birds may descend up to 100 feet (30 m). Normally they stay under for less than 30 seconds, swimming 20–30 feet (6–9 m) below the surface.

DOUBLE-CRESTED CORMORANT

CLASS **Aves**

ORDER **Pelecaniformes**

FAMILY **Phalacrocoracidae**

GENUS AND SPECIES **Phalacrocorax auritus**

ALTERNATIVE NAME
White-crested cormorant

WEIGHT
3.5–10 lb. (1.5–4.5 kg)

LENGTH
**Head to tail: 30–36 in. (75–90 cm);
wingspan: about 55 in. (1.4 m)**

DISTINCTIVE FEATURES
**Slender, streamlined body; long neck;
long, sharply hooked bill; large webbed feet.
Adult: mainly glossy black plumage with
orange-yellow facial skin and throat pouch;
ear tufts in breeding season; juvenile: light
brown plumage, palest on neck and breast.**

DIET
Almost exclusively fish; some crustaceans

BREEDING
**Age at first breeding: 3 years; breeding
season: usually eggs laid April–June;
number of eggs: usually 2 to 4; incubation
period: about 25–30 days; fledging period:
about 42 days; breeding interval: 1 year**

LIFE SPAN
Probably up to 20 years

HABITAT
**Sheltered marine waters, especially near
coasts; inland lakes, rivers and swamps**

DISTRIBUTION
**Western and eastern coastlines of North
America; Caribbean; parts of central
Canada, Great Lakes area and Midwest**

STATUS
Common in most of range

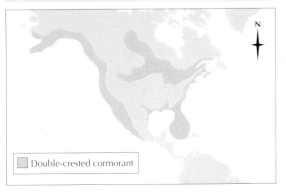

Double-crested cormorant

Fishing in flocks

Cormorants feed mainly on fish, occasionally supplemented with a few crustaceans such as crabs. The flightless cormorant feeds on octopuses and fish, mainly eels, which it finds in great numbers around its island home, where the cool Humboldt Current wells up and mixes with warm equatorial waters to form a fertile zone abundant in plankton and fish.

The great cormorant and the shag, *Phalacrocorax aristotelis*, have slightly different diets. The great cormorant feeds mainly on bottom-living sea animals including flat fish and sand eels, while the shag eats eels and fish from the middle and upper waters. In this way the two closely related species, which nest on the same cliffs and hunt in the same waters, do not compete for food.

Cormorants often hunt together in flocks, though they also fish alone. The birds can be seen flying in lines then settling on the water, gathering in a tight bunch and lowering their heads into the water to look for fish. Flocks of some species are reported to locate shoals of fish then swim round them in decreasing circles to draw the prey together, and double-crested cormorants drive shoals of fish up a bay into shallower water. Fish are brought to the water's surface, to

The flightless cormorant has lost the ability to fly. It no longer has strong flight muscles and its tiny wings are useful only as paddles.

The king cormorant, like most species of cormorant, nests in crowded colonies. It is a native of the Falkland Islands in the South Atlantic.

be swallowed headfirst. Large fish are first shaken or beaten against the water until they cease to struggle.

Relationship with humans
Japanese fishers train cormorants to bring fish back to their boat; leather collars are used to prevent the birds from swallowing the catch. However, in many other parts of the world fishers persecute cormorants because of their voracious appetite for fish. In some regions cormorants' appetites have given rise to an important industry. The guanay cormorant, *P. bougainvilli*, of Peru and Chile, and the Cape cormorant, *P. capensis*, of South Africa, feed in vast numbers, and their droppings, deposited on the breeding grounds, form guano. This is dug out and used as a rich fertilizer. In Walvis Bay, Namibia, special platforms have been built to attract cormorants for this purpose.

Breeding colonies
Cormorants nest in colonies, which vary from a few pairs to thousands of pairs, depending on the species and location. The colonies are usually situated on rocky cliffs and the nests may be within a few feet of each other. Other colonies, especially those inland, may be situated in dead trees. The nests are usually a bowl of twigs, grasses, reeds and seaweed.

Courtship displays involve much waving of the birds' long necks; the courtship rituals of the flightless cormorant may take place in the water. When soliciting a mate, the female bends her neck over her back. The eggs, usually two to four in number, are incubated by both parents. The chicks hatch in 3–5 weeks and at first are naked,

with skins like black leather. Later they grow a curly, dark gray down. The parents feed their chicks on fish, which the youngsters take by pushing their heads down the parents' gullets. The chicks usually leave the nest in 5–8 weeks.

Keeping their distance
At rest, a cormorant often perches with its wings held fully outspread. The usual explanation for this behavior is that it is related to the cormorant's habit of flapping its wings to dry off after immersion in the water. It has been suggested that this is necessary because the cormorant's wings are not well waterproofed, but this would be most surprising in a family of birds that are not only aquatic, but have relatives such as pelicans, boobies and gannets that are also aquatic. No other aquatic birds, except darters, use this method to dry out their wings and, furthermore, close observation does not suggest that standing with wings spread is connected with drying. Cormorants can be seen holding their wings open in pouring rain, when drying is impossible, or after they have flown from one perch to another, when drying is not necessary. On other occasions they may leave the sea and perch with wings folded.

Cormorants habitually perch a short distance from each other, so that a group of the birds on a rock are evenly spaced out in a line. When a new bird lands and joins the group, it extends its wings and its neighbors shift away. The newcomer folds its wings and the line of cormorants is still well-spaced with a wingspan between each one. This would seem to suggest that wing-spreading is a device adopted to keep individual cormorants apart.

CORN BORER

THE CORN BORER IS a native of Europe, where it is sometimes known as the maize moth, and was accidentally introduced to North America in the early 1900s. The American name is more commonly used because it is in the United States that the species has become a serious pest of corn, or maize as the crop is called in Europe. The adult corn borer is a small, inconspicuous moth with a wingspan of about 1 inch (2.5 cm). The female is yellowish brown with dark, wavy lines running irregularly across the wings; these markings are often referred to as zigzags. The male is very much darker in color with markings of olive brown.

Colonization of North America

The larvae of corn borers have been found on about 250 species of plant, including corn, beets, celery, beans, garden flowers, potatoes, peppers and tomatoes, although corn is by far the most frequently attacked crop. In England the larvae have also been found on hops.

It is thought that the first corn borers to reach North America may have arrived in cargoes of corn from Hungary or Italy. The first positive discovery of this pest in the United States came in 1917, when crops near Boston, Massachusetts, were found to be infected. Two years later the corn borer appeared in the Great Lakes region. Before long the species had become so well established that large-scale programs of eradication failed, and the insect eventually reached the Midwest cornbelt.

Today the corn borer is one of the most serious agricultural pests in the United States. Occasionally entire crops of corn are ruined because the larvae weaken the stems to such an extent by burrowing that the plants topple over. Corn borer larvae not only leave corn plants physically weakened but also increase their susceptibility to disease. If modern insecticides had been available in 1917 the insect's spread might have been checked but the compulsory burning of corn stubble and weeds after harvesting proved to have little effect.

Destructive life cycle

The eggs of the corn borer are laid on the undersides of leaves in groups of 20 to 30, each female laying up to 600. They hatch in 5–15 days,

The larva of the corn borer, a small and rather dull-colored moth found in much of North America and Europe, is a serious pest of corn.

depending on the temperature. The larvae have dark heads and pale yellow to brown bodies, and have a smooth texture. They bear several rows of small blackish spots and become darker just before hatching.

Corn borer larvae are responsible for the crop damage for which the species has become infamous. They usually begin feeding on leaf surfaces, chewing small, round holes in them. Thereafter the larvae start boring in the mid-ribs of the leaves until they reach ¾–1 inch (2–2.5 cm) in length. Shortly after hatching, the young larvae disperse over several plants. During dispersal and prior to boring into plants they are more susceptible to natural predators and to adverse weather and cold temperatures.

Mature larvae overwinter inside tunnels in corn stubble or in the ears of corn or other protective plant material. They pupate in spring and adult moths emerge during April and May. A second generation may hatch out in the summer, but it is more usual for them to hatch during the spring. Once the adult moths have emerged, they mate. Eggs are laid and a new generation is born. Adult corn borers are mostly active in late evening and during the night. By day they hide in weeds, among grass and in any trash found in the field.

Corn the perfect host

Scientists have discovered that one reason the corn borer favors the corn plant as a host is that the life cycle of corn corresponds well with the moth's own life cycle. Corn is planted in early spring and the leaves are sprouting just in time for the moths to lay their eggs. The leaves provide a steady source of food through the year and in winter the dead stalks provide a very necessary shelter for the resting larvae.

Closer examination revealed that the larvae prefer some parts of the corn plant to others. At first the larvae feed in the tightly rolled whorl of leaves wrapped around the stalk. When the flower head develops, they feed there. As the flowers expand, the larvae move back inside the leaf bases and into the husks that surround the ears. Tests have shown that this movement is due to their dislike of light and to their preference for being in crevices with as much of the body as possible touching something solid.

Even within these distinct parts of the plant, the corn borer larvae exercise preferences. They move to flowers rather than leaves, and to inner husks rather than outer husks because these parts contain more sugar. Young corn plants contain a chemical that is toxic to corn borers, and sugar acts as an antidote to the poison. The chemical is the corn's natural remedy against corn borers. However, although some of the

larvae are killed by the toxin, those that consume a sufficient amount of sugar are safe. Research is being conducted into the feasibility of breeding corn varieties resistant to the larvae and of introducing natural predators to reduce populations. The biological control agents may include bacteria, such as *Bacillus thuringiensis*, nematode worms, such as *Steinernema carpocapsae*, and parasitic wasps, such as *Eriborus terebrans*.

CORN BORER

PHYLUM **Arthropoda**
CLASS **Insecta**
ORDER **Lepidoptera**
FAMILY **Pyralidae**
GENUS AND SPECIES ***Ostrinia nubilalis***

ALTERNATIVE NAMES
European corn borer; maize moth

LENGTH
Adult wingspan: about ¾–1¼ in. (2–3 cm). Larva: newly hatched, 1.5 mm; fully grown, about 1 in. (2.5 cm).

DISTINCTIVE FEATURES
Adult female: pale yellow to light brown in color; outer third of wings usually crossed by dark zigzag lines. Adult male: darker and more slender than female; zigzags often pale yellow. Larva: pale brownish body with several rows of black spots; very dark head.

DIET
Adult: does not feed. Larva: about 250 known plant hosts; favors corn but also attacks beans, beets, hops, celery, potatoes, peppers and tomatoes.

BREEDING
Breeding season: spring and summer; number of eggs: 500 to 600; hatching period: 5–15 days, depending on temperature; larval period: up to about 20 days; breeding interval: 1 generation per year in cooler environments, 3 or 4 generations per year in warmer climates such as southern states of U.S.

DISTRIBUTION
Native to most of Europe; introduced to North America in early 20th century and now found in much of Canada and U.S.

STATUS
Superabundant in agricultural areas when conditions are ideal

CORNCRAKE

ALSO CALLED THE LAND RAIL, the corncrake belongs to the rail family, Rallidae. It is related to the coots and gallinules, which it resembles in form, though it is slightly smaller. The corncrake spends more time on land than other rails, which are primarily wading birds. Its plumage is yellowish buff and marked with black on the upperparts; the head and breast are grayish. The wings are a rusty red color, and are conspicuous in flight.

Farmland birds

The corncrake breeds in Europe and central Asia, from southern parts of Scandinavia and Siberia south to the Mediterranean coast and northern China. It is absent from Spain, Italy, Greece and the Balkans. During years of warm weather the corncrake extends its range northward but, overall, its range contracted dramatically during the 20th century. Agriculture helped the spread of the corncrake at first, for the species is mostly a grassland bird, but over time modern farming methods contributed to its decline. For example, in the early 1900s corncrakes were common in much of the British Isles. Now they are found only in the wilder parts of western Ireland and of the Hebrides and Orkney Islands, which lie to the north and west of Scotland, all areas where there is little intensive agriculture.

Corncrakes feed mainly on insects such as beetles, earwigs, weevils and flies, including the eggs and larvae as well as the adults. Slugs, snails, earthworms, millipedes and spiders are eaten in smaller quantities, while plant food taken includes seeds and some greenery.

Winters in Africa

The corncrake is strongly migratory, wintering in southern Asia and sub-Saharan Africa, as far south as South Africa. Asian corncrakes have been known to reach Australia on very rare occasions. When flushed unexpectedly, corncrakes fly weakly, with legs dangling, but when the birds are migrating, their flight is steadier. Corncrakes fly near to the ground and another possible cause of their decline is that they often hit overhead power lines and cables.

Even where the corncrake is common, it is rarely seen. The bird is mostly active during the evening, when it keeps to the cover of vegetation, taking off only when alarmed. However, its presence is well advertised by the two-beat call

of the male, which is produced for hours on end by day and night. The sound is often described as a creaking and the species' common and scientific names are derived from this call.

Caring for the young

Migrant corncrakes reach the British Isles from their winter quarters in mid-April to mid-May. The males begin to call a few days after their arrival and continue to do so for 2–3 weeks, only stopping when the eggs are laid. They may start calling again after hatching and can be heard until late August, about a month before they begin their migration back south.

The male displays to the female with head held low, wings spread so the tips just touch the ground, and with the feathers of the neck and sides fluffed out to form a ruff. He circles around her, moving his head from side to side, and she turns all the time to keep facing him until mating commences. The eggs, up to 14 in number, are laid in a flat pad of plucked grass among grasses, nettles or low undergrowth. Low-lying water meadows seem to be preferred, but corncrakes are also found in upland areas.

The eggs are incubated by the female for 16–19 days. As she does not start incubation until the last egg is laid, all of her chicks hatch out

Corncrakes are very secretive birds, rarely seen by humans. They hide among tall grasses and are most active in the evening and early morning.

The corncrake's scientific name, Crex crex, *is derived from the distinctive, far-carrying call of the male.*

within 24 hours. For about 4 days the chicks are fed by the female or by both parents. After this period they run about with the parents and feed themselves. The chicks start to fly when they are about 5 weeks old, although they cannot fly well for a further 2–3 weeks.

The corncrake and mechanization

Populations of the corncrake are thought to have declined by as much as 50 percent over 20 years during the second half of the 20th century. The species is classed as vulnerable by the I.U.C.N. (World Conservation Union), and in the 1990s it was estimated that there were about 165,000 breeding pairs left in Europe.

The corncrake's downfall was brought about by the introduction of mowing machines and the progressively earlier start to the the hay-cutting season, with the result that the date for hay-making coincided with the species' nesting season. This led to the destruction of eggs and young chicks. In Holland, corncrakes were to some extent able to avoid this fate by moving from grasslands into fields of corn, but this behavior offered only a temporary respite because there has been a similar trend for the early harvesting of cereal crops.

In Britain, corncrake numbers began to fall in the late 19th century. The decline started in the southeast and spread northward and westward, though in the early 20th century corncrakes could still be heard near London. The last significant British corncrake populations began to disappear after World War II, when agricultural mechanization reached the Scottish islands and western Ireland. Farmers try to avoid corncrake nests but these are well camouflaged and cannot always be seen in time. The birds' best chance of survival is when bad weather delays the harvest.

CORNCRAKE

CLASS	**Aves**
ORDER	**Gruiformes**
FAMILY	**Rallidae**
GENUS AND SPECIES	***Crex crex***

ALTERNATIVE NAME
Land rail

WEIGHT
Up to ½ lb. (225 g); male larger than female

LENGTH
Head to tail: 10½–12 in. (27–30 cm); wingspan: 17–21 in. (43–53 cm)

DISTINCTIVE FEATURES
Laterally compressed body; long legs and toes; short tail; mainly yellow brown with chestnut wings, reddish brown bars on side of body and undertail and dark streaks on upperparts; female duller than male

DIET
Mainly beetles, earwigs, weevils, flies and other invertebrates; some seeds and shoots

BREEDING
Age at first breeding: 1 year; breeding season: mid-April–July; number of eggs: usually 8 to 12; incubation period: 16–19 days; fledging period: 34–38 days; breeding interval: 1 year

LIFE SPAN
Probably up to 15 years

HABITAT
Areas of long grass, including unimproved pasture, alpine meadows and marshes

DISTRIBUTION
Breeds in parts of Europe, east through Central Asia to northern China; winters in eastern Africa and (probably) southern Asia

STATUS
Vulnerable; large population declines throughout Europe

Corncrake (breeding range)

COTINGA

THE NAME COTINGA COMES from an Amazon Indian word meaning "washed white" and originally described the white bellbird, or snowy cotinga, *Procnias alba*. A near relative is the three-wattled bellbird, *P. tricarunculata*, which has red-brown plumage with a white head and chest and (in the male) three whiplike wattles dangling from the base of the bill.

There are about 90 species in the cotinga family, Cotingidae, which differ so widely in form that they were once placed in several families. Even now, the relationship of different cotingas to one another is a matter of some debate among scientists. Many of the species in the cotinga family include the word cotinga as part of their English name; among the other species are the fruiteaters, purpletufts, tityras, pihas, becards, fruitcrows and umbrellabirds.

Many male cotingas have ornate plumage with brilliant colors, as well as crests, wattles and other adornments. Only the males are showy; the females tend to be drab and inconspicuous. The three most ornate cotingas are the Andean and Guianan cock-of-the-rock (*Rupicola peruviana* and *R. rupicola*) and the umbrellabirds (in the genus *Cephalopterus*). Male cock-of-the-rock are intense red or orange with a unique sail-like crest, while the three species of umbrellabird have black, umbrella-like crests and long wattles. Some species of cotinga, such as the tityras and the becards, are relatively plain, although the white-winged becard, *Pachyramphus polychopterus*, is very striking in spite of its dull color. The male is black on the upperparts with a gray rump and white wing patches. The female is olive-green above and pale yellow below.

Fruits of the forest

Cotingas evolved in the dense forests of the Amazon River basin, but over time various species spread south to the northern borders of Argentina and north through Central America to the southern borders of the United States. Today cotingas are found in lowland primary forest, montane forest and rain forest; the greatest diversity of species is present in the canopy. One species, the Jamaican becard, *Pachyramphus niger*, has reached the Caribbean, where it lives only in the highlands of Jamaica.

Many cotingas are exclusively fruit eaters, though some species also take a few invertebrates. The tityras, which have stout, hooked bills, catch dragonflies and lizards, and white-winged becards have been observed picking insects off foliage while in midair.

Mysterious voices

There are almost as many lifestyles as there are body forms among the cotingas. Many species live in the upper layers of foliage in the treetops, but others occur in clearings or at the forest edge. Cotingas are best known for their persistent calling. The bellbirds, for instance, like the unrelated birds of Australia and New Zealand that bear the same name, produce loud, bell-like peals that can be heard a mile or so away. The calfbird, or capuchinbird, *Perissocephalus tricolor*, makes a mooing or grunting call. The tityras also "grunt" in a distinctive manner. Other cotingas are less vocal, and some have rather quiet songs.

Females rear young alone

Male cotingas often perform elaborate courtship rituals. One of the most spectacular is that of the cock-of-the-rock, in which groups of males dance together at traditional display arenas. The male bearded bellbird, *Procnias averano*, courts and mates on a special display branch, as does the capuchinbird, which clears its chosen branch of twigs by snapping them off in its bill.

After courtship and mating most female cotingas rear their families on their own. This is true even of tityras, which remain paired all year; the male tityra does not help with nest-building

Many male cotingas are brilliantly colored, including the male blue cotinga, Cotinga amabilis, *of Venezuela, Colombia and Ecuador.*

BEARDED BELLBIRD

CLASS   **Aves**

ORDER   **Passeriformes**

FAMILY   **Cotingidae**

GENUS AND SPECIES   *Procnias averano*

ALTERNATIVE NAMES
Mossy-throated bellbird; black-winged bellbird

LENGTH
Head to tail: 5½–7 in. (14–18 cm)

DISTINCTIVE FEATURES
Male: whitish plumage with coffee-brown head and black wings; long, black stringlike wattles hanging from bare throat. Female and young male: olive-green upperparts and yellowish underparts; no throat wattles.

DIET
Forest fruits

BREEDING
Age at first breeding: 3 years; breeding season: mainly April–October; number of eggs: 1; incubation period: 23 days; fledging period: about 33 days; breeding interval: 1 year

HABITAT
Canopy of dense tropical forests

DISTRIBUTION
Northeastern Columbia east through northern Venezuela; southeastern Venezuela north to northwestern Guyana

STATUS
Probably uncommon

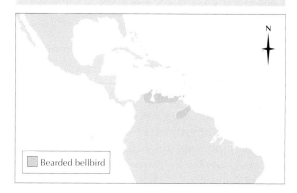

Bearded bellbird

Cotingas are typically solitary, secretive birds of thick forest. The swallow-tailed cotinga, Phibalura flavirostris, feeds and nests high in the treetops.

or with incubation of the two eggs. Similarly, the male white-winged becard sings to his mate while she takes care of construction of the nest. This is a bulky construction, built high in the treetops with an entrance in the side. The female masked tityra, *Tityra semifasciata*, makes her nest in a hole, usually one abandoned by a wood-pecker. The nest is created by dropping leaves and twigs into the cavity.

Competition over nests

A good nest site is a valuable resource for a bird and suitable sites may be in short supply in forests that have large bird populations. Certain parasitic flycatchers have been known to destroy clutches of becard eggs and to rear their own families in the empty nests. Moreover, not all hole-nesting birds are able to excavate their own, and such species often use cavities abandoned by other birds, particularly woodpeckers. Tityras go even further: they oust the original inhabitants by harassing them until they are forced to move. At first the tityras may occupy the nest-hole while the owner is absent. They then start to bring nesting material and drop it into the cavity. The birds that first had the hole remove the leaf litter that the tityras bring; however, the tityras generally seem to win ownership of the nest through sheer persistence.

Family under threat

The rain forests of Central and South America are rich in birds but are increasingly threatened by large-scale clearance to create land for agriculture, roads, dams, mines and human settlement.

The cotingas are one of the most threatened bird families in the region, and the I.U.C.N. currently lists 15 species of cotingas as threatened. The kinglet calyptura, *Calyptura cristata*, is critically endangered, and the banded cotinga, *Cotinga maculata*, is endangered; the remaining 13 species, which include two of the three species of umbrellabirds, are all considered vulnerable.

COTTONTAIL

THE COTTONTAILS ARE SMALL rabbits that vary in color from dark gray to red-brown. The upperparts are brown and the flanks are gray, with a rufous nape and legs. The ears are relatively short and rounded. The tail is brown above and white below and has given rise to the rabbits' common name because of its resemblance to a ball of cotton wool. There are 13 species of cottontail rabbits, the head and body length of which is 8–20 inches (20–50 cm), according to species; the largest species may weigh ten times as much as the smallest.

Thrive almost anywhere

Cottontail rabbits range from central Canada south to South America, as far south as Argentina and Paraguay. Over this range they live in a very wide variety of habitats. Most species prefer open woodland and brush or clearings in forests. Consequently cottontails flourished in early colonial times when European settlers were first opening up the forests; later, complete destruction of forests had a detrimental effect on certain species. The New England cottontail, *Sylvilagus transitionalis*, of the Appalachian Mountains, prefers open woodland with plenty of under-

growth. As a result of the spread of agriculture this species has been largely replaced by the eastern cottontail, *S. floridanus*, which is more adaptable. Today the eastern cottontail is the best known rabbit in eastern North America. It is abundant in most habitats within its range, including open country, farmland, swampy woods, thickets and scrub.

Specialized cottontails

Some species of cottontails are adapted to life in more extreme habitats. The smallest species, the desert cottontail, *S. audubonii*, lives in arid grasslands and semideserts. Nuttall's cottontail, *S. nuttallii*, favors arid woodlands and sagebrush as well as montane uplands. The largest species of cottontail, the swamp rabbit, *S. aquaticus*, which is sometimes called the canecutter, lives in marshes and other wetlands. It has large feet with splayed, slightly furred toes, and swims well. When alarmed it is said to make for water, where it hides with only nose and ears visible.

Cottontails are timid animals, ready to bolt for cover if they sense danger. Having found a suitable hiding place the rabbits crouch motionless, their fur coats blending in so well with the

Cottontails are grazers and browsers with a varied diet that includes many types of vegetation. Their habit of feeding on seedlings and uprooting young crops makes them unpopular wih farmers.

Nuttall's cottontail is widespread in the Rocky Mountains, from southernmost Canada south to New Mexico.

COTTONTAILS

CLASS	**Mammalia**
ORDER	**Lagomorpha**
FAMILY	**Leporidae**
GENUS	***Sylvilagus***

SPECIES   **13, including eastern cottontail, *S. floridanus*; Nuttall's cottontail, *S. nuttallii*; desert cottontail, *S. audubonii*; New England cottontail, *S. transitionalis*; swamp rabbit, *S. aquaticus*; and brush rabbit, *S. bachmani***

LENGTH
Head and body: 8–20 in. (20–50 cm), depending on species

DISTINCTIVE FEATURES
Prominent, rounded ears; white "bobbed" tail; large eyes; coat generally reddish brown or gray brown but varies according to species

DIET
Grasses, shoots, buds, leaves, twigs and bark

BREEDING
Age at first breeding: 11–12 weeks; breeding season: spring until fall; number of young: usually 5; gestation period: about 30 days; breeding interval: usually 3 litters per year

LIFE SPAN
Up to 5 years

HABITAT
Varies according to species; includes brush, scrub, farmland, prairies, marshes, forests, mountains and arid semideserts

DISTRIBUTION
Central Canada south to South America

STATUS
Most species generally common; critically endangered: Omilteme cottontail, *S. insonus*; endangered: Dice's cottontail, *S. dicei*; Tres Marias cottontail, *S. graysoni*

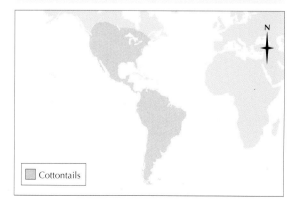

Cottontails

surrounding vegetation that they can be difficult to find. Most cottontail species are crepuscular (active during twilight hours). However, in the summer months, when nights are short, cottontails are more active and are more likely to be seen in broad daylight.

Individual territories

Cottontails do not generally stray far from their birthplace. The female usually spends the whole of her life within a territory measuring no more than a few hundred square meters. The male cottontail ranges over a wider area, which usually includes the territories of several females. Territorial boundaries are strictly observed. If chased, a cottontail will refrain from entering another's territory for as long as possible, running in circles until exhausted, at which point it will take refuge in any available burrow.

Each cottontail has a range of several acres, which is crossed by regular runs through the ground vegetation. The animal knows its way so well in these runs that it can move at a top speed of 20–25 miles per hour (30–40 km/h) when frightened. A cottontail uses the abandoned burrow of another animal, such as a groundhog or prairie dog, for refuge and as a winter shelter. However, most females of breeding age dig new holes in which to give birth.

Double digestion

Cottontails eat grasses, herbs and broad-leaved annual plants and can severely damage crops. Some crop plants are trampled and others only nibbled, inflicting sufficient damage to spoil them as a food crop. Young plants treated in this way grow stunted and deformed. In winter cottontails feed mainly on buds and soft twigs, and young saplings can be found neatly cut off at the level of snowdrifts.

Digestion of coarse herbage and twigs is difficult for cottontails, as such materials contain large quantities of indigestible cellulose, which has to be assimilated. Other grazing animals, such as cows and sheep, assimilate their food by chewing the cud. In that process food initially passes to the rumen, the first compartment of the multiple stomach, where the cellulose is broken down; the food is then brought up to the mouth and taken back to the true stomach. In cottontail rabbits another process is used: food is passed through the digestive system twice, to ensure complete digestion.

The pelletlike droppings of cottontail rabbits come in two forms: soft and green, and hard and brown. The soft, green pellets are eaten after being expelled, a process known as coprophagy. This process enables the rabbits to assimilate any undigested food still present in the droppings. The green pellets pass through the digestive system, remaining nutrients are extracted and hard, brown pellets are excreted.

Long breeding season

Cottontail rabbits typically have a lengthy breeding season, which lasts from February to late September in temperate parts of North America. In arid regions with a hot climate, such as the Sierra Nevada, breeding starts earlier in the year because by summer the vegetation will have shriveled and there will be little food for the young. Breeding may take place all year round in regions where conditions are favorable.

Fighting sometimes occurs between rival males and courtship rituals may take place in which one rabbit leaps in the air while the other runs under it. After mating the male is driven away and the female rears the litter alone. The young cottontails are born about a month later, naked and blind. There may be up to 12 of them,

The desert cottontail has relatively large, thinly furred ears that facilitate heat loss in a hot environment.

Female cottontails produce litters of up to 12 young, although five is normal, and have three or more litters each season.

weighing less than 1 ounce (28 g) each. The usual litter size is three to seven. The young are placed in a shallow nest that the mother has scraped in the ground and lined with fur plucked from her breast. When the female leaves the nest she covers it with grasses in order to keep the young warm and to hide them from predators. The nest is too small for the mother to lie in herself. Instead, she crouches over it and the babies have to climb up to suckle.

When the young are 2 weeks old they leave the nest and feed nearby during the day. They finally disperse after 4–5 weeks. Their mother will have mated again within a few hours of their birth and the next litter will soon be due. A female cottontail will have three, or occasionally up to five, litters per year.

Prey for many animals

Cottontails fall prey to the majority of the large predatory animals sharing their environment. Skunks, foxes, coyotes and crows kill the young rabbits in their nests, while owls, hawks and snakes take the young when they leave. Other predators, such as bobcats and eagles, take the adult cottontails. Humans hunt and trap cottontails and in the eastern United States they are the chief type of game; sometimes several million cottontails are killed annually in a single state. The fur is of little use except in the manufacture of felt, but they are good for the table.

One method of studying an animal's dietary preferences is to examine the undigested remains passed in its feces. Cottontails form the staple food of many American predators, in much the same way that the European rabbit is the mainstay of many carnivores and birds of prey on the other side of the Atlantic. In a study made in the Sierra Nevada, cottontails were found to be an important prey item of coyotes, gray foxes, bobcats, horned owls and gopher snakes. According to the study about 60 percent of the horned owl's diet in this region appears to consist of cottontail rabbits. However, cottontail populations are usually able to sustain great losses. The rabbits have a rapid reproductive rate and, if food is abundant and the weather favorable, their populations soon increase dramatically.

COUCAL

DESPITE BEING MEMBERS OF the cuckoo family, Cuculidae, coucals lay their eggs in their own nests rather than using the nests of other birds. There are 28 species of coucal and the birds are rather large in comparison to other cuckoos. Some authorities include the couas of Madagascar in the same subfamily (Centropodinae) as the coucals. The couas are very similar to the coucals both in form and habit, but there are only about 10 species.

The pheasant coucal, *Centropus phasianus*, of Australasia, is usually about 24 inches (60 cm) in length; its long tail, characteristic of the cuckoo family, gives the species a pheasant-like appearance. In the breeding season the adult male pheasant coucal is glossy black on the head and underparts, with rich brown wings and a greenish black tail and rump, which are barred with brown and white. The adult female is similar, but larger and with an orange-yellow iris, which is bright scarlet in the male.

The common coucal, *C. sinensis*, of Southeast Asia, is sometimes known as the crow-pheasant. As with the Australasian pheasant coucal, its common name refers to the long, pheasant-like tail feathers. In India and Malaysia, which are home to many true pheasants, the common coucal's large size mean that it can be misidentified as one of those species as it runs through undergrowth or flies heavily across open spaces.

Shy and retiring
Coucals are found in sub-Saharan Africa, from Senegal east to Somalia and south to the Cape, as well as in Asia and Australasia, ranging as far east as the Solomon Islands. Although widespread in some places, the coucals' secretive habits means that they are not well known. Many coucals, including the pheasant coucal and the greater and lesser black coucals (*C. menbeki* and *C. bernsteini* respectively, both native to New Guinea), live in swampy country. They prefer to skulk in thick cover, flying as little as possible; the Senegal coucal, *C. senegalensis*, hops from bough to bough through bushes. An exception to this general rule is provided by the blue-headed coucal, *C. monachus*, which can often be seen in the morning and evening when it sits in the tops of reeds and high grasses.

Being so secretive in their habits, coucals are best known by their calls. Many have "bubbling" calls that sound like water being poured from a bottle. Others have a whooping call, from which the name coucal is derived. The blue-headed coucal utters a low *cou-cou-cou* that is immedi-

Coucals are more often heard than seen because they live in densely vegetated habitats and can move unobtrusively through thick foliage. The Senegal coucal (above) is one of the most familiar species.

ately answered by its mate; the call and answer being used to keep in touch while moving about in thick undergrowth. The white-browed coucal, *C. superciliosus*, has a "bubble" call; a pair will call together, one bird calling at a higher pitch. This bird also has a harsh *chak* call, most likely to be heard during and after rain; as a result the white-browed coucal is referred to as the rainbird in some parts of Africa.

The blue-headed coucal has the most broadly based diet of any coucal. It feeds on insects, hunts small birds, lizards, reptiles, mice and rats and is also a scavenger. The white-browed coucal, which sometimes lives in gardens and fields, takes large numbers of grasshoppers and snails as well as snakes and other vertebrates. The pheasant coucal steals from the nests of other birds, taking eggs and nestlings, and occasionally takes to raiding chicken runs. However, insects and spiders make up the bulk of its diet.

Well-hidden nests
Coucal nests are well camouflaged and built near the ground, in a low bush or tussock of grass. Coucals usually choose sites among swamps, wet heathlands or vegetation bordering watercourses. The pheasant coucal's nest is built in a large tussock by drawing the tops of the grass stems together to make a hollow 4 inches (12 cm)

in diameter, and is lined with green leaves. The nest has an opening at either end and the coucal sits on the eggs with its head and long tail protruding. Sometimes these nests have an entrance tunnel, also lined with green leaves, which are replaced as they dry out and turn brown. Pheasant coucals have also been known to use old nests deserted by babblers.

Both sexes incubate the three to five eggs for about 2 weeks. The chicks take about 17 days to fledge, and at first bear a long, white hairlike covering; the developing feathers are contained within horny sheaths. The white-browed coucal is reported to carry its young, one at a time, between its thighs when danger, such as a forest fire, threatens. This behavior is well known in various species of woodcock, and it is likely that when all the coucals have been studied other species of coucal will be found to do the same.

Threatened species

Several species of coucals are now at risk, due primarily to the clearing of their natural habitat by humans. Destruction of forest cover on the island of Sri Lanka has led directly to a decline in the population of the green-billed coucal, *C. chlororhynchus*, and the species is currently regarded as endangered. The black-hooded coucal, *C. steerii*, inhabits lowland forest on the island of Mindoro in the Philippines. As this habitat is now scarce and is likely to be completely cleared within the next 10–20 years, this species is faced with extinction.

The coppery-tailed coucal, Centropus cupreicaudus, is common throughout South Africa. It frequents thick vegetation near rivers and water meadows.

PHEASANT COUCAL

CLASS	**Aves**
ORDER	**Cuculiformes**
FAMILY	**Cuculidae**
GENUS AND SPECIES	***Centropus phasianus***

ALTERNATIVE NAMES
Common coucal; swamp cuckoo

WEIGHT
12–17½ oz. (340–500 g); female slightly larger than male

LENGTH
Head to tail: 21–31½ in. (53–80 cm)

DISTINCTIVE FEATURES
Slender body; strong, decurved bill; very long, black tail; relatively short legs; underparts glossy black (breeding plumage) or yellowish (nonbreeding plumage); chestnut-brown wings; iris bright red (breeding male), orange yellow (breeding female) or brown (juvenile)

DIET
Spiders, insects, snails, bird eggs and nestlings, mud crabs and frogs

BREEDING
Age at first breeding: 1 year; breeding season: eggs usually laid September–May; number of eggs: 3 to 5; incubation period: 15 days; fledging period: 17 days; breeding interval: up to 3 broods per year

LIFE SPAN
Not known

HABITAT
Dense riverside vegetation, swamps, damp grasslands, thickets and overgrown gardens

DISTRIBUTION
New Guinea; Timor and East Timor; coastal regions of northern and eastern Australia

STATUS
Fairly common

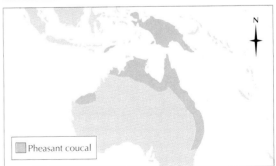

Pheasant coucal

COURSER

OGETHER WITH THE PRATINCOLES, the nine species of courser form a family of long-legged, plover-like shorebirds. The species most familiar to ornithologists is the cream-colored courser, *Cursorius cursor*, a blackbird-sized bird. It is a pale sandy color, with creamy legs, black primary wing feathers and a distinctive, black and white eye stripe. Temminck's courser (*C. temminckii*), Burchell's courser (*C. rufus*) and the Indian courser (*C. coromandelicus*) look rather similar, although the plumage of these species is much darker than that of the cream-colored courser. The Egyptian plover, *Pluvianus aegyptus*, is also a courser, but it differs from the other species in its appearance: the upperparts are pale gray while the underparts are pale yellow or whitish; a white stripe runs across the head, the rest of which is black.

Coursers range from Africa, where seven of the nine species breed, to southern Asia. The cream-colored courser breeds from Morocco east to India and south to Kenya. It is common around the fringes of the Sahara and occasionally wanders north into Europe. Of the two Indian species, Jerdon's courser, *Rhinoptilus bitorquatus*, was presumed to have become extinct until 1986, when it was rediscovered.

Desert runners

Coursers generally live in dry, sandy or stony places and in grassland in arid regions. The exceptions are the Egyptian plover, which occurs along sandy riverbanks, and Temminck's courser, which is found in open spaces in woods and forests. As their name suggests, coursers are good runners; like many birds that have specialized in running, they have lost the fourth, backward-facing toe. Coursers can fly well and some species migrate considerable distances, but they prefer to avoid danger on foot rather than by taking flight. They also have the habit of stopping to stretch up on tiptoe, craning their necks to peer around them, and when alarmed often simply crouch down to conceal themselves against the ground. Coursers live mainly on insects such as beetles, grasshoppers, ants, flies and butterfly larvae, though some species also take snails and small lizards.

Chicks protected from the sun

As a rule coursers do not make a nest, and lay their two or three eggs on the bare ground, but Temminck's courser sometimes makes a shallow depression for them. The eggs are incubated, or shielded from the sun, by the female alone. The

Pale plumage provides the cream-colored courser with excellent camouflage in the arid lands of Africa and western Asia.

Egyptian plover broods its eggs during the night when the air and ground cool down rapidly, while in the daytime it covers the eggs with sand to prevent them from getting too hot.

The young can run soon after hatching, but are shaded from the sun by each parent in turn during the middle of the day. Another habit characteristic of the Egyptian plover and, it is presumed, of other coursers, is to hastily bury the chicks if danger threatens. The chicks flatten themselves against the ground, if possible in a small depression; hippopotamus footprints have been described as being a suitable hiding place. The parent courser then kicks sand over the crouching chicks and makes its own escape, running away and luring the enemy from the young. Chicks have been found 1 inch (2.5 cm) or more under the surface, which is sufficient to protect them from the worst of the heat, while not so deep that they are prevented from breathing. The parents cool the sand by fetching water from a nearby river and regurgitating it over the place where the chicks are buried.

Rediscovery of a species

Jerdon's courser was originally known as an inhabitant of the Penner and Godaveri valleys in eastern and central India. Its preferred habitat is sparse, thorny scrub on uneven, rocky ground. Assumed extinct for most of the 20th century, Jerdon's courser was rediscovered in January 1986. The species has since been found at six sites in the vicinity of the Lankamalai mountain ranges, near the Penner valley. At one point this area was threatened by a proposed irrigation scheme, but two protected zones have now been created. Despite these moves, Jerdon's courser is still listed by the I.U.C.N. (World Conservation Union) as endangered.

The double-banded courser, Rhinoptilus africanus, is found on bare, stony terrain in southern Africa. Unlike other coursers, it lays only one egg.

EGYPTIAN PLOVER

CLASS	**Aves**
ORDER	**Charadriiformes**
FAMILY	**Glareolidae**
GENUS AND SPECIES	***Pluvianus aegyptus***

ALTERNATIVE NAME
Crocodile bird

WEIGHT
2½–3 oz. (75–90 g)

LENGTH
**Head to tail: 7½–8⅓ in. (19–21 cm);
wingspan: 18½–20 in. (47–50 cm)**

DISTINCTIVE FEATURES
Long legs and wings; black and white markings on head, neck and back; pale gray upperparts; buff-colored underparts

DIET
Insects and other invertebrates

BREEDING
Breeding season: January–April; number of eggs: usually 1 to 3; incubation period: 28–31 days; fledging period: 30–35 days; breeding interval: 1 year

LIFE SPAN
Not known

HABITAT
Sandy areas beside rivers; occasionally also in fields and near human settlements

DISTRIBUTION
Much of tropical sub-Saharan Africa

STATUS
Uncommon

Egyptian plover

COWBIRD

THE THREE GENERA OF COWBIRDS are related to grackles, troupials and American blackbirds. Cowbirds are fairly small birds, with typically dark, glossy plumage. One of the most common species is the shiny cowbird, *Molothrus bonariensis*, which has a conical, finchlike bill and glossy greenish-black and violet plumage. The screaming cowbird, *M. rufoaxillaris*, has a mainly blue-black plumage with some green in the wings and tail. The male brown-headed cowbird, *M. ater*, is greenish black apart from its brown head; the female of the species is gray brown all over. In common with most cowbirds, all three species are parasitic, laying their eggs in the nests of other bird species. Parasitic cowbirds do not construct their own nests or care for their own young, which are instead raised by the owners of the host nests.

The cowbirds are confined to the Americas, most species being found in South America. Some have a relatively restricted range. For example, the bay-winged cowbird, *M. badius*, is found in parts of Argentina, Bolivia, Paraguay and Uruguay, and on the coast of northeastern Brazil. Other species are more widespread. The shiny cowbird is found over a considerable range

in South America, from Venezuela and Colombia south to the mid-region of Argentina. The brown-headed cowbird occurs across the continent of North America, from the Pacific coast to the Atlantic coast. The northern boundary of its range runs from Nova Scotia, along the northern shores of the Great Lakes, up to the Great Slave Lake and down to Vancouver; the southern boundary is in Mexico.

Opportunistic feeders

Cowbirds are so called because of their habit of following cattle to feed on insects flushed from the grass by the animals' large hoofs. They live in open country, spending the winter months in flocks and splitting into pairs during the breeding season. Some species are sedentary; others, such as the shiny and brown-headed cowbirds, migrate in the spring and fall.

Cowbirds feed mainly on seeds, fruits and berries; insects and other invertebrates are also taken, though mostly during the summer when they are abundant. Many kinds of seeds are eaten and some species of cowbirds will feed on spilt grain around farmyards. In some regions cowbirds are a considerable menace to fields of

Most cowbirds do not build their own nests but lay their eggs in the nests of other species. This nestling cowbird is being reared by a pair of yellow warblers, Dendroica petechia, alongside the warblers' own young.

More than 250 species of birds are parasitized by the brown-headed cowbird (male, above). Its favorite hosts are other members of the family Icteridae.

BROWN-HEADED COWBIRD

CLASS	**Aves**
ORDER	**Passeriformes**
FAMILY	**Icteridae**
GENUS AND SPECIES	***Molothrus ater***

ALTERNATIVE NAME
Buffalo bird (archaic)

WEIGHT
1¼–2 oz. (35–60 g)

LENGTH
**Head to tail: 6¾–8⅓ in. (17–21 cm);
wingspan: 11½–13¾ in. (29–35 cm)**

DISTINCTIVE FEATURES
**Stout, conical bill; dark, glossy plumage.
Male: coffee-brown head and metallic,
green-black body; female: dark gray brown
with streaks on underparts; juvenile: closely
resembles female but paler and more
heavily streaked.**

DIET
**Seeds including rice, corn and other crops;
berries; invertebrates, especially in summer**

BREEDING
**Breeding season: March–July; number of
eggs: 10 to 12; incubation period: 11–12
days (in host nest); fledging period:
21 days (in host nest); breeding interval:
1 year (female lays variable number of
eggs in several host nests)**

LIFE SPAN
Up to 7 years

HABITAT
**Prairies, farmland, woodland edge,
hedgerows, gardens and city parks**

DISTRIBUTION
Central Canada south to Mexico

STATUS
Common; increasing in north of range

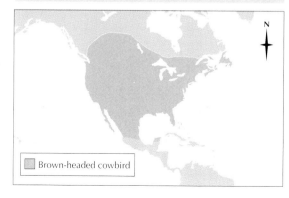

Brown-headed cowbird

growing crops. This is the case particularly in the late summer when breeding is over and cowbirds descend on the ripening fields of rice or corn in large, dense flocks.

Cowbirds may also be beneficial to farmers, however, in that they feed on the grubs that are thrown up in the wake of plows. They were once known as buffalo birds in the prairies of North America; when large herds of buffalo (American bison) were still in existence, the cowbirds associated with them, walking behind the buffalo as they grazed, or perching on their backs.

Parasitic life cycle

The parasitic habit of laying eggs in another bird's nest is not restricted to certain species of cuckoo. Several groups of birds exhibit this behavior, including the cowbirds. More than 250 different species are parasitized by the brown-headed cowbird, although some, including orioles, warblers and sparrows, are preferred to others. Most brown-headed cowbird eggs are laid in the nests of its relatives, the orioles.

Cowbirds can have a significant effect on the breeding success rate of the species that they parasitize. One study revealed that up to 80 percent of all nests belonging to the song sparrow, *Melospiza melodia*, are used by cowbirds of one species or another. Some cowbirds are highly host-specific while others, such as the shiny and brown-headed cowbirds, are less restricted in their choice of host. These two species regularly parasitize the nests of other members of the blackbird family, including orioles, caciques and oropendolas.

A female cowbird lays her eggs in a host nest at the same time that the owner is laying hers, and before that species has started incubation. By keeping a close watch, the female cowbird notes when the chosen bird leaves her nest, then she enters and lays an egg of her own in less than a minute. In this way the cowbird usually avoids detection. Some time later, seizing another convenient opportunity, the female cowbird returns to the host nest to remove one or more of the host's eggs in her bill.

Each female cowbird lays a variable number of eggs, usually one in each of several host nests, but if there is a shortage of suitable nests several cowbirds may share the same nest. As many as 37 shiny cowbird eggs have been found in the nest of a single ovenbird, *Seiurus aurocapillus*.

Some birds are able to recognize and reject cowbird eggs. Tyrant flycatchers react to the threat of parasitism by building a new floor to their nest, which covers the alien, trespassing eggs; the covered eggs cool and never hatch. The gray catbird, *Dumetella carolinensis*, and American robin, *Turdus migratorius*, actively throw cowbird eggs out of their nests. However, some birds react to parasitism by deserting their nests altogether and then nesting again elsewhere.

If a cowbird egg is accepted the future nestling stands an excellent chance of survival. Cowbird eggs hatch in a shorter time than the eggs of the host birds, so cowbird nestlings have a good start in the competition for food that exists between nestlings. Sometimes this results in the death of the other species, but not invariably, and two or three other nestlings, as well as the cowbird, may be successfully reared. The young cowbird grows fast and leaves the nest in under 2 weeks, by which time it is able to fly.

Nonparasitic cowbirds

The bay-winged cowbird rears its own offspring, both parents helping to feed the young. Very often the bay-winged cowbird builds its own nest out of grasses and feathers, most of the work being done by the male. At other times it occupies the abandoned nests of other species.

Widespread in South America, the shiny cowbird extended its range northward during the 20th century and is now also found in the Caribbean.

COWRIE

The shell of a living cowrie (mole cowrie, Cypraea talpa, above) is completely encased by a protective organ called the mantle, which often features large protuberances.

THE COWRIE IS A MOLLUSK belonging to the same group as the sea snails, but with a shell that shows nothing of the external spiral structure familiar in those animals. The shell whorls laid down in the early life of a cowrie become enveloped during growth until only the final whorl is visible. On the underside of the mature shell is a long slitlike opening with its sides rolling inward, like a scroll. The margins of the opening are grooved or toothed. The surface of the shell is usually smooth and glossy, although in some of the smaller cowries it may be grooved. When the cowrie is living the shell is usually covered by the folds of the part of the mollusk's body known as the mantle. These meet at the midline of the upper surface, and help to protect the shell. The mantle is often ornamented with pointed or forked filaments.

Most cowries are found on the coral shores of the Tropics, in the Indian and Pacific oceans, north almost to Japan and south to New South Wales in Australia. They also occur in parts of the Mediterranean and along both coastlines of North America. The largest cowries are several inches long. In temperate seas there are far fewer species, the shells of which are relatively small and less ornate.

Cowries were once associated with the worship of the goddess Aphrodite, which was inaugurated in Cyprus, hence the generic name *Cypraea*. The name cowrie is a corruption of "gowry" or "kauri," from Hindi and Urdu respectively, and seems to have entered the English language in the mid-17th century. The small European cowrie, *C. europaea*, is traditionally known as the "nun" in some localities and as "maiden stick-farthing" or "grotty buckie" in others.

Carnivorous lifestyle

Sea snails feed mainly on seaweed, rasping pieces from the plants with a filelike tongue, or radula, but cowries are an exception: most species are carnivorous. Their staple food is small anemones, sponges, compound ascidians (sea squirts), coral polyps, dead mollusks and the eggs and egg-capsules of other sea snails. The mouth is at the end of a tubelike proboscis, through which the radula is protruded.

A few cowrie species have been found in waters as deep as 1,650 feet (500 m), but most are shallow-water mollusks and live between the tidemarks. The lifestyle of cowries is similar to that of other mollusks, and consists of alternate resting periods and feeding sorties. When active a cowrie creeps over the seabed, its shell hidden by the mantle flaps, on which a number of small eyes are situated. In many cowries there is a pair of eyes in front, one near the base of each of the two sensory tentacles, which are extended forward when the cowrie is moving. Cowries quickly withdraw their tentacles at the slightest disturbance in the water.

Egg-laying and larval growth

Reproduction in the various species of cowrie differs slightly in the way the eggs are laid. The European cowrie makes a small hole in the jellylike bodies of sea squirts, into which it deposits a vase-shaped capsule, about ¼ inch (6 cm) high, containing several hundred bright yellow eggs. One cowrie will deposit a number of such capsules over a short period of time. Fertilization is internal and sperm is stored, each batch of eggs being fertilized as it is laid.

The chestnut cowrie, *C. spadicea*, of the Pacific coast of the United States, is about 1½ inches (4 cm) long and in July lays its eggs in rounded capsules. These are pointed at the apex, and each contains 800 eggs. The capsules are laid in tightly packed batches of about 100, forming a roughly circular plate nearly 2 inches (5 cm)

COWRIES

PHYLUM	Mollusca

CLASS **Gastropoda**

ORDER **Mesogastropoda**

FAMILY **Cypraeidae**

GENUS ***Cypraea, Trivia*, many others**

SPECIES **Several hundred, including European cowrie, *C. europaea*; money cowrie, *C. moneta*; and chestnut cowrie, *C. spadicea***

LENGTH
Most species: ⅔–3 in. (1.5–8 cm)

DISTINCTIVE FEATURES
Domed shell with long, slitlike opening on underside and (in many species) pointed or forked filaments. Shell surface usually smooth and glossy, often with striking coloration and markings.

DIET
Wide variety of sedentary animals, including sponges, coral polyps, anemones and sea squirts; also detritis and seaweed (some species only)

BREEDING
Age at first breeding: usually 1–3 years; number of eggs: typically several hundred; larval period: varies according to environment and species

LIFE SPAN
Not known

HABITAT
Mainly shallow coastal waters, especially on coral reefs; few species in waters up to 1,650 ft. (500 m) deep

DISTRIBUTION
Most species in tropical seas; some species in subtropical and warm temperate seas

STATUS
Some species common

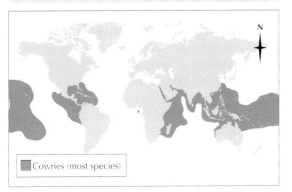
Cowries (most species)

Cowries (Cypraea chinensis, above) are able to withdraw their tentacles and mantle at speed. Their coloration abruptly changes from that of the mantle to that of the shell, startling predators.

in diameter. The plate is guarded for 3 weeks by the cowrie, which expands its foot into a circular sheet and allows this to hover overhead.

The larvae hatching within a capsule make their way out through an opening in the top of the capsule and are free-swimming for a while. Then each begins to grow a shell, descends to the seabed and grows into a young cowrie, initially with a spiral shell. Gradually the spiral shell is overgrown by the last whorl, after which growth ceases, whereas in other mollusks growth is continuous throughout life.

In cowries the central column, or columella, of the early spiral shell is dissolved and absorbed by the mollusk and the chalky material used to lay down the final shell. Shell-forming marine animals must store considerable amounts of calcium in their body tissues prior to an addition to their shells. Calcium is present in the sea in relatively very small quantities but marine animals are highly efficient at extracting it. They cannot afford to waste any and the material must be absorbed and re-used when possible.

Cowrie currency

The most widespread cowrie in the Indian Ocean is the money cowrie, *Cypraea moneta*. Its shells have been used in trade from very early times until recently in parts of the Malay Archipelago, the Maldive Islands and other areas fringing the Indian Ocean. They have even been carried to Scandinavia and to parts of North America, where Native Americans once traded with them.

In West Africa cowries were a medium of exchange until the mid-19th century, and their use as currency spread inland to the heart of Africa as far as Timbuktu. In coastal areas the shells were threaded in strings of 40 or 100, 50 of the first and 20 of the second then being worth about one American dollar. This is a considerable bulk of shells for such a small sum of cash.

COYOTE

THE COYOTE IS CLASSIFIED with wolves and domestic dogs in the genus *Canis*. The name coyote comes from the Mexican *coyotl* and can be pronounced with or without the "e" silent. A coyote weighs 15–50 pounds (7–20 kg) and measures about 4 feet (1.2 m) from nose to tail-tip. The fur is gray or tawny and the tail, bushy with a black tip, droops low behind the hind legs, rather than being carried horizontally in the manner of wolves. The coyote is smaller and hunts smaller game than the gray wolf, though it is larger than the red fox.

Coyotes once lived on plains and in woods to the west of North America; they were known as brush wolves in forested regions and as prairie wolves in open lands. In the 19th century the species' range began to increase despite persecution from humans and coyotes are now one of the most widespread mammals in the Americas, ranging from northern Alaska south to Panama. They have also spread eastward to the Atlantic seaboard. By the early 1900s coyotes had reached Michigan; they were seen in New York State in 1925 and in Massachusetts in 1957. Coyotes had spread to the southern shores of Hudson Bay by 1961. They are now found in most areas of Canada except for the extreme north.

Unlike most predatory mammals found in North America, the coyote expanded its range during the 19th and 20th centuries. It occurs from Canada south to Panama in a wide variety of habitats, from prairies to city suburbs.

An adaptable species

In the face of persistent harassment by humans most carnivorous animals have retreated. Their habitat has been destroyed and they are hunted mercilessly as vermin or as valuable fur bearers. In contrast, the coyote is extending its range. There is no market for its fur but the coyote has long been shot on sight, or trapped and poisoned, because it has been regarded as an enemy of livestock and competes with humans for game. According to one estimate about 125,000 coyotes have been killed annually for many years, yet the species continues to flourish. The coyote's powers of survival seem to lie in its wariness, its ability to adapt to new circumstances and its broad diet. Generally a solitary animal, it also hunts in packs, which are capable of bringing down larger prey, such as sheep.

The spread of the coyote into the northeastern United States is probably linked with widespread tree-felling and with the regional decline of the gray wolf. The demise of that species left a niche that the coyote was able to fill. Even urban development has not deterred the coyote. It has moved into suburbs where, like the red fox in Britain, it can supplement its diet with gleanings from trash cans and other sources.

COYOTE

CLASS	**Mammalia**
ORDER	**Carnivora**
FAMILY	**Canidae**
GENUS AND SPECIES	***Canis latrans***

ALTERNATIVE NAMES
Brush wolf; prairie wolf (both archaic)

WEIGHT
**Male: 20–50 lb. (9–20 kg);
female: 15–40 lb. (7–18 kg)**

LENGTH
**Head and body: 2½–3 ft. (75–90 cm);
tail: 1–1⅓ ft. (30–40 cm)**

DISTINCTIVE FEATURES
**Resembles gray wolf, but smaller, lighter
and has more slender muzzle; long coat,
usually gray on upper side and paler below**

DIET
**Mainly small mammals, especially jack
rabbits, cottontail rabbits and rodents; some
birds, fish, insects, fruits, nuts and carrion;
occasionally poultry, sheep, goats and deer**

BREEDING
**Age at first breeding: 1–2 years; breeding
season: usually January–March; number of
young: usually 6; gestation period: 58–65
days; breeding interval: 1 year**

LIFE SPAN
Up to 15 years, usually much less

HABITAT
**Varied, including woodland, grassland,
prairie, tundra, chaparral and scrub**

DISTRIBUTION
**Alaska east to Nova Scotia and south to
Panama; range extending southward**

STATUS
Very common

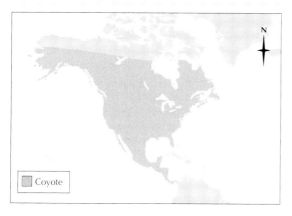
Coyote

The coyote is the only wild member of the dog family that often barks, although it also has a range of other calls, the best known of which is its distinctive howl.

Persecuted as a pest

Coyotes are persecuted because of their reputation as killers of livestock and deer. While sheep, goats and deer are occasionally killed, the coyotes' reputation is largely undeserved and has probably arisen due to the commonplace sightings of coyotes feeding on carrion.

In studies to establish whether coyotes pose a major threat to livestock, several thousand dead specimens were examined. Their stomachs were found to contain mainly jack rabbits and cottontails, together with mice, voles and other small rodents. Poultry and livestock made up

Play among coyote pups helps to hone hunting skills that will be needed as adults.

about one-eighth of the sample. It is probable that, as with other animals that have a varied diet, coyotes will eat whatever is most readily available. If rabbits are abundant, then poultry runs are left alone; however, if a weak calf is found, a coyote will attack it. Coyotes feed on a range of animals: insects, birds, trout and cray-fish have all been found in coyote stomachs. They sometimes attack and eat beavers, domestic cats, skunks and even gray foxes. At one time coyotes were a major cause of mortality in the swift, or kit, fox, *Vulpes velox*; coyotes are also known to take red foxes and the rare San Joaquin kit fox, *V. v. mutica*. Sometimes coyotes eat large amounts of vegetable matter, including prickly pears, grasses, nuts and ripe water melons.

Coyotes usually hunt singly or in pairs, and are capable of attaining speeds of over 40 mph (65 km/h). Sometimes they chase deer in relays, one coyote taking over the pursuit as another becomes tired. Another habit is to play dead, waiting for inquisitive, carrion-eating birds such as crows to land and examine the "corpse," at which point the coyote attacks them.

Rearing the young

Breeding begins when coyotes are a year old; a male and female pair for life. They mate during January to March and the pups are born 58–65 days later. The den is usually made in a burrow abandoned by a groundhog, skunk or fox, and is enlarged to form a tunnel up to 30 feet (10 m) long and 1–2 feet (30–60 cm) in diameter, ending

in a nesting chamber that is kept scrupulously clean. In some circumstances, for example in marshlands where tunnels would be flooded, nests are sometimes made on the surface.

Up to 12 pups may be born in a litter, the average number being around six. They are born with their eyes shut and stay underground for over 1 month. The father stays with the family, bringing food first for the mother, then for the pups. The mother regurgitates the food to her young in a partly digested form. Later the family go out on communal hunting trips and the pups learn to hunt for themselves, finally leaving the parents when they are 6–9 months old. Although coyotes are efficient hunters, they are by no means immune to attacks from larger predators, and are known to have been killed by gray wolves and pumas. Golden eagles will some-times attack young coyotes.

The call of the coyote

The coyote is classified as *Canis latrans*, literally "barking dog," because apart from the domestic dog it is the only member of the dog family that habitually barks. Foxes, wolves and jackals bark only at specific times. Coyotes can be heard all the year round, usually at dawn and dusk. In the evening coyotes sing in chorus. An individual starts with a series of short barks, gradually increasing in volume until they merge into a long yell. Other coyotes join in and the chorus continues for a minute or two. After a pause, the chorus starts again.

COYPU

THE COYPU IS A LARGE RODENT that is ratlike in appearance and measures over 3 feet (1 m) from the large, square muzzle to the tip of the long, scaly tail. The ears are short and the small eyes are set high in the large face. The coypu's large, orange-colored incisors help it to chew through the tough water vegetation that forms the major part of its diet.

The coypu is adapted for an aquatic life, and is a skilled swimmer and diver; its natural habitats include swamps, marshes, canals and ponds. It is equipped with webbed hind feet and two layers of fur: the long, coarse guard hairs effectively protect the soft, waterproof underfur. The coypu's underfur is called nutria by furriers, a name that is derived from the Spanish word for otter, and which is sometimes used instead of the name coypu.

Reared for their fur

Coypus are native to central and southern South America where they are widely distributed in swamps and along watercourses. During the 19th century they were extensively hunted for their fur and numbers dropped until laws were passed for the species' protection. Numbers rose again, but crashed when an epidemic struck. By this time, coypus, like other fur-bearing animals such as chinchillas and mink, were being exported from their native habitats and reared on farms. Like other fur bearers, they frequently escaped and went wild in their new environments. Today there are wild populations of coypus living in many areas of northern and western Europe, central Asia and Japan.

The species is also widespread in the United States. Coypus are particularly abundant in the marshlands of Louisiana, which they colonized from a population of 20 individuals brought to the state in 1938. In the late 1950s there were an estimated 20 million coypus in Louisiana; by the early 1960s they were the state's leading fur-bearing animal.

In England coypus were first found wild shortly after fur farms were established in the 1930s. Before and during World War II they were released in large numbers as the farms closed down. The coypus colonized rivers and marshes, especially in the Norfolk Broads in southeastern England, and their numbers rapidly increased.

Coypus look a little like giant rats with their large muzzles, long whiskers, prominent incisor teeth and long, scaly tails.

As their populations grew, coypus began to attack crops and there were fears that their burrows, normally above the waterline, would cause extensive flooding when water levels were very high. Accordingly, a campaign to exterminate the animals was started in 1962. At the time the coypus were widespread in the east of the country. The policy was to contain the spread of the animals, then drive them back. It was successful mainly because the hard winter of 1962–1963 killed off most of the population. Today the breeding population in the British Isles is confined to the Norfolk Broads, with only stragglers being found elsewhere.

Coypus live in pairs or family groups amid dense aquatic vegetation where they make runs and build nests. Occasionally large colonies form, consisting of several small groups living in close proximity to one another. When food is plentiful, for example in farming areas during the summer, coypus are predominantly active during the evening and at night, spending the whole day resting in their burrows. During the winter, coypus sometimes emerge in the daylight hours to forage. In the summer, when thick vegetation is readily available, coypus may build themselves nests in the open. However, coypus usually prefer to live in burrows underground.

As in many aquatic animals, the coypu's eyes, ears and nostrils are positioned toward the top of its head. It can therefore see, hear and breathe while it is almost entirely submerged.

COYPU

CLASS	**Mammalia**
ORDER	**Rodentia**
FAMILY	**Capromyidae**
GENUS AND SPECIES	***Myocastor coypu***

ALTERNATIVE NAME
Nutria

WEIGHT
Up to 37½ lb. (17 kg), usually 11–22 lb. (5–10 kg)

LENGTH
Head and body: 17–25 in. (43–64 cm); tail: 10–17 in. (25–43 cm)

DISTINCTIVE FEATURES
Resembles large rat; large, square muzzle; small eyes; webbed hind feet; long, scaly tail; large, orange-colored incisor teeth

DIET
Aquatic plants, especially rushes, sedges, reed mace, grasses and weeds; also crops, mollusks and small fish

BREEDING
Age at first breeding: 3–7 months; breeding season: all year; number of young: usually 4; gestation period: 128–130 days; breeding interval: 2 or 3 litters per year

LIFE SPAN
Up to 2–3 years

HABITAT
Swamps, marshes, lakes, ponds, canals and slow-flowing streams

DISTRIBUTION
Native to southern Brazil, Bolivia, Paraguay, Uruguay, Argentina and Chile; introduced to parts of U.S., Canada, northwestern Europe, Central Asia and Japan

STATUS
Common in native and introduced ranges

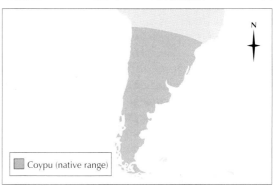

Coypu (native range)

They may build their own holes or take over the holes of other animals, and the tunnels they build may be simple or may form a complex system, perhaps extending 50 feet (15 m) or more. This activity goes some way toward explaining the coypu's reputation as a destroyer of riverbanks and dikes.

Preference for coarse food

The parts of plants that the coypu seems to prefer are not always the most succulent. It appears to favor the tissues around the base of plants, including the roots. Rushes, sedges, reed mace, Canadian water weed, water grasses and water parsnip are all part of the species' diet and most of these plants are fibrous. As the coypu's diet consists of a high percentage of aquatic plants that grow at the bottom of pools and lakes, the animal must dive deeply and use its teeth and forepaws to uproot the plants. As the coypu cannot chew and swallow underwater, it brings the plants to the surface and eats them on the bank. A large amount of debris may build up as a result of feeding on the bank, and coypus use these platforms to rest on or to shelter on during bad weather.

Coypus antagonize farmers when they move away from riverbanks and attack fields of crops, notably sugar beet. The animals also attack other root crops and cabbages and pull down wheat to get at the grain. Coypus supplement their vegetarian diet with small fish and freshwater mollusks, such as snails and mussels.

Babies suckled in the water

In England, coypus breed throughout the year, raising two to three litters of up to 13 young, after a gestation of 128–130 days. The babies are well-developed at birth. Their eyes are open and they have full coats of fur; within a few hours they can move around, and when a few days old they begin taking solid food.

The mother's nipples are on the sides of her body so that the babies can feed while she lies on her stomach. More importantly, the nipples are above the surface when she swims, so she can suckle her litter without coming ashore. Coypus are weaned at 2 months and are able to breed 1 month later, before they are fully grown.

The coypu's enemies

In South America, jaguars, caimans and other large carnivores feed on coypus, but in England the adults have no natural predators. Normally docile, adult coypus can defend themselves with their sharp incisor teeth. Young coypus fall prey to a wide range of animals, including ermines, otters, rats, hawks and owls.

The coypu's outer layer of bristlelike fur protects an inner layer of dense, soft underfur that helps to waterproof the animal.

CRABEATER SEAL

The crabeater seal is probably the world's most abundant seal, but the inaccessibility of its pack ice habitat has made an accurate population count impossible.

OF THE FIVE TRUE SEAL SPECIES living around the coasts of the Antarctic, the crabeater is the smallest. It is a slender, lithe animal with a small head. The adults of both sexes measure at most 8–9 feet (2–2.5 m) from snout to tail and weigh 440–660 pounds (200–300 kg). Due to their creamy white fur, crabeaters are also known as white seals. The fur molts in January (midsummer in the southern hemisphere) and the seals take on a grayish brown appearance. During the year this fades back to white. As crabeater seals age their fur generally becomes lighter all year round.

Life in the pack ice

Crabeaters are the most numerous seals in the Antarctic, if not the world; their population has been estimated at up to 15 million. However, accurate figures are difficult to establish because the seals live in pack ice and can be reached only by icebreaker or by airplane. Crabeaters are noticeably more common than other seals, but the number of seals seen basking on the ice floes is not a true indication of their real numbers: it is impossible to know how many other seals are in the water. This number depends, at least in part, on the weather, for seals prefer to come out in calm, sunny conditions. In summer crabeater seals tend to move south as the pack ice breaks up, and are found closer to the shore; more detailed studies may show these movements to be regular migrations from north to south. Some individuals wander north, reaching Australia, New Zealand and Uruguay.

Crabeater seals can move over the ice at a surprisingly fast speed. They throw themselves forward by pushing at the packed snow and ice with hind and front flippers in an undulating, caterpillar-like action, and have been recorded moving at 15 miles per hour (24 km/h).

Unusual diet for a seal

Like most seals the crabeater seal hunts fish, but its main food is the shrimp-like crustacean called krill, the same animal that forms the staple diet of penguins, whales and many Antarctic seabirds. To catch these crustaceans in sufficient numbers to be efficient, the crabeater seal has a device to strain them from the water that resembles the baleen plates of some whales. In the crabeater seal the teeth act as a strainer and each tooth has four or five cusps. The seal sucks in a mouthful of krill and water, and shuts its mouth so the teeth on the upper and lower jaws fit together. It then forces water out through the gaps between the cusps, retaining the krill. The seal's straining system is completed by bony growths at the back of the jaws that fill the gap behind the teeth present in other mammals.

The crabeater seal may have the most specialized teeth of all the carnivores. The peculiar shape of its teeth is probably derived from the sharp, cusped teeth of species such as the leopard seal, an arrangement used by those species to grip slippery fish and squid.

Pups born on ice floes

Female crabeater seals are thought to mate first when they are 2 years old, bearing their single pups a year later, in October. Although crabeaters often gather on the ice in large numbers, pupping takes place in small groups. The seals prefer ice where the floes are not too closely packed together and have been thrown into hummocks that provide shelter from the wind. The newborn pups are about 4 feet (1.2 m) long, 55 pounds (25 kg) in weight and are covered in a soft, woolly coat. Scientific opinion differs as to whether the pups are suckled for 2 or 4 weeks. The baby coat is presumably shed when the pup is weaned and takes to the sea to fend for itself. Crabeaters probably live for up to 30 years.

CRABEATER SEAL

CLASS	**Mammalia**
ORDER	**Pinnipedia**
FAMILY	**Phocidae**
GENUS AND SPECIES	***Lobodon carcinophagus***

ALTERNATIVE NAME
White seal

WEIGHT
440–660 lb. (200–300 kg)

LENGTH
Head and body: 8–9 ft. (2–2.5 m)

DISTINCTIVE FEATURES
Slender body; small head; light cream-white coat, becoming gray brown in summer

DIET
Mainly krill; some fish

BREEDING
Age at first breeding: 2–3 years; breeding season: October–December; number of young: 1; gestation period: about 270 days; breeding interval: not known

LIFE SPAN
Up to 30 years

HABITAT
Cold seas and pack ice

DISTRIBUTION
Antarctica and surrounding seas

STATUS
Abundant; population estimated to be at least 15 million

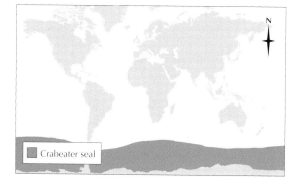

Crabeater seal

Scarred by other seals

Before anything else was known of the crabeater seal's life, explorers in the Antarctic had noticed that the seals sometimes bore huge scars running the length of their bodies. It seems surprising that the seals are able to survive such injuries until it is realized that only the blubber and

surface muscles are cut open, and that the seals therefore suffer no serious damage from their injuries. It was once thought that these scars were the legacies of attack by killer whales but examination of a killer whale's mouth shows that its teeth could not fit the scars. It is now believed that the scars are from bites by leopard seals, incurred when the crabeaters are young.

Crabeater mysteries

The survey base at Hope Bay is at the northern tip of the Antarctic Peninsula; just to the south of it is the Crown Prince Gustav Channel, which separates James Ross Island from the mainland. For most of the year the channel is frozen over and survey parties find it a convenient route for their sledging trips. In the winter of 1955 they found vast concentrations of crabeater seals on the ice. A total of 3,000 seals, 10 times more than usual, was counted in various parts of the channel. More surprisingly, 2 months later, in mid-October, most of these seals had died. In one group only 3 percent of seals survived.

Samples of the seals' internal organs were preserved and sent back to England where it was found that a disease had been the cause of death. A virulent epidemic had clearly spread through the population, decimating seal numbers in a manner similar to outbreaks of myxomatosis in the European rabbit. However, the mystery disease did not affect the entire population of crabeater seals. It was an isolated incident, and until there are further outbreaks its cause will remain a mystery.

The crabeater seal has a misleading name. It feeds mainly on tiny, shrimplike animals. Its highly specialized teeth act as a strainer, enabling it to expel mouthfuls of sea water and retain the prey.

CRAB-EATING FOX

ALSO CALLED THE CRAB-EATING dog, the crab-eating fox is classified as a member of the dog family, Canidae, though it is neither a true dog nor a true fox. The English language has evolved only three generic names for members of the dog family: "dog," "fox" and "wolf;" terms that do not have a strictly scientific meaning. These words have been used to name many canids purely on the basis of the animals' size or appearance. Among the species named in this way are several animals native to South America, such as the bushdog, the now extinct Antarctic wolf of the Falkland Islands, and the crab-eating fox.

The crab-eating fox is the common canid of Colombia and Venezuela, and ranges south from these countries to Peru, southeastern Bolivia, northern Argentina and Paraguay. It is found in tropical and subtropical woodland on both mountains and plains, and also along the banks of rivers, inhabiting areas of scattered trees as well as denser forest. In some regions, seasonal movements occur, and the animals retreat to higher ground during the wet season.

It has been traditionally believed that the crab-eating fox hunts in parties of five or six, but this figure is probably based on observations of family groups. The crab-eating fox is in fact usually solitary, sometimes hunting in pairs. It is active mainly at night, and spends the day in the abandoned burrow of another animal.

Inaccurately named

In spite of its name, the crab-eating fox does not specialize in hunting crabs. Its diet consists mainly of small mammals and birds, which it hunts by scent in woodland or runs down in open country. Insects, mostly grasshoppers, frogs and lizards, are also eaten. The crab-eating fox takes fruit as well, including figs, berries and cultivated fruits, such as bananas. When living near rivers the crab-eating fox catches freshwater crabs, but it also eats mollusks and other animals that live in the shallows. On the plains it digs out turtle eggs.

Early naturalists were responsible for giving the crab-eating fox its common name, and at one time it was known scientifically as *Canis cancrivorus*, from the Latin cancer, meaning crab. These scientists affirmed that crabs were the main item of the animal's diet and that small mammals and birds, now known to

The crab-eating fox is nocturnal and, despite its name, has a varied diet that includes mammals, birds and insects.

The crab-eating fox resembles the more familiar red and gray foxes in its sharp muzzle, pointed ears and bushy tail. Its fur is short and varies in color, ranging from pale gray to dark brown, often with yellowish-brown or black coloration in places. The head and body are about 2 feet (60 cm) long and the tail is 1 foot (30 cm) in length; the ears have a dark tip. The fur of the crab-eating fox is less valued than that of other South American dogs and foxes because of its short length. In Buenos Aires this species' fur is marketed as "Brazil fox" or "provincial fox."

be the main food, were eaten only when crabs were not available. The fox was reputed to dive into shallow water for crabs, though authoritative scientific works make no mention of this habit. Considering the name of the fox, scientists would be unlikely to overlook any such observations, especially as members of the dog family are not usually good at swimming and diving.

It seems likely that early naturalists observed the crab-eating fox on the banks of rivers, which formed the best, if not the only, routes through the unopened South American continent. The naturalists saw that the animals fed on crabs.

CRAB-EATING FOX

CLASS	**Mammalia**
ORDER	**Carnivora**
FAMILY	**Canidae**
GENUS AND SPECIES	***Cerdocyon thous***

ALTERNATIVE NAMES
Crab-eating dog; crab-eating zorro; wood fox

WEIGHT
13–17½ lb. (6–8 kg)

LENGTH
Head and body: about 2 ft. (60 cm); tail: 1 ft. (30 cm)

DISTINCTIVE FEATURES
Short, broad ears; broad head; bushy tail; short-haired, gray-brown coat

DIET
Mainly small mammals (especially rodents) and birds; also domestic fowl, lizards, frogs, invertebrates, iguana and turtle eggs, fruits and carrion

BREEDING
Age at first breeding: 9–12 months; breeding season: all year; number of young: 3 to 6; gestation period: 52–59 days; breeding interval: at least 8 months

LIFE SPAN
Probably up to 12 years

HABITAT
Forest, light woodland and open grasslands in both mountains and lowlands

DISTRIBUTION
Northern Colombia and Venezuela south through Brazil to Paraguay and northern Argentina; absent from much of Amazon River Basin

STATUS
Probably common throughout range

Crab-eating fox

Based on inadequate knowledge, the animals were thus given an inaccurate common name. Indeed, the species' full scientific title incorporates the names of three different animals: *Cerdocyon thous* comes from the Greek words *kerdo* ("fox"), *cyon* ("dog") and *thous* ("jackal").

The crab-eating fox is generally solitary, but sometimes it hunts in pairs and when breeding can be seen in family groups.

Breeding

Little information is available on the breeding habits of the crab-eating fox, other than that gained by observations of captive animals. Crab-eating foxes form stable, monogamous pair bonds, in which both parents assist with guarding the den and with caring for pups. In captivity, adult females can come into estrus at any time of the year, a phenomenon known as aseasonal estrus. Some females come into heat twice annually. Gestation lasts for about 56 days, though there is considerable variation. Litters usually consist of three to six pups, which disperse from the natal den at about 6–8 months of age. The young reach sexual maturity at the end of their first year.

CRAB PLOVER

The crab plover is a unique shorebird placed in a family of its own. It is found on coasts in the Indian Ocean and Red Sea.

THE CRAB PLOVER IS PLACED in a family of its own, Dromadidae, as it is distinct from other shorebirds both in its form and its habits. An unusual shorebird, about 13–14 inches long (33–36 cm), the crab plover has a conspicuous black and white plumage. Most of its body is pure white but the flight feathers and the back are jet black; as a result, the crab plover resembles the Eurasian avocet, *Recurvirostra avosetta*. Young birds have a grayish brown plumage. The legs are long and dark in color and the toes are partially webbed. The tail is short and the large black bill is long, somewhat flattened and gull-like, having a pronounced gonys (a ridge formed by the junction of the two halves of the lower mandible near its tip).

Crab plovers are found on sandy and muddy shores and on coral reefs around the coasts of the Indian Ocean, from Natal in South Africa east as far as the Andaman Islands in the eastern Indian Ocean, including in the Red Sea and The Gulf. The species is now known to be partially migratory, with a very restricted breeding range that includes parts of the southern Red Sea, Gulf of Aden and Gulf of Oman. Crab plovers are visitors only in the southern and eastern parts of their wide range, flying there after breeding.

Sociable birds

Crab plovers are noisy, gregarious birds. Outside the breeding season they are found in flocks and have been seen perching on hippos as they lie in the water. The flocks are typically made up of 20 to 30 birds; winter flocks may number 150 to 200 in some areas, and, exceptionally, up to 400. The birds fly or stand in tight formations, searching for food in the mud and on the shore, and making chattering barks and screeches that can be heard some distance away.

Feed on the strandline

Crab plovers feed at the top end of the intertidal zone, seeking out marine animals that become stranded and exposed by the ebbing tide. As their name suggests, the birds' main prey is crabs. These are caught in the bill and beaten against the ground until they are unable to escape. Small crabs are swallowed whole and large ones are ripped apart with the crab plover's powerful bill and eaten piecemeal. Crab plovers are usually nocturnal, as the fiddler crabs on which they often feed emerge at dusk to forage and mate. Worms, crustaceans, mudskippers and fish spawn are also eaten, as are shellfish, which are first battered to crack them open.

CRAB PLOVER

CLASS	**Aves**
ORDER	**Charadriiformes**
FAMILY	**Dromadidae**
GENUS AND SPECIES	***Dromas ardeola***

WEIGHT
8–11½ oz. (230–325 g); female heavier

LENGTH
**Head to tail: 13–14 in. (33–36 cm);
wingspan: 30–31 in. (75–78 cm)**

DISTINCTIVE FEATURES
**Large, gull-like bill; long, pointed wings;
long, dark-colored legs with partially
webbed feet; short tail. Adult: mainly white
plumage with black areas on wings and
back; juvenile: grayish brown plumage.**

DIET
**Mainly crabs; also other marine crustaceans,
worms, mudskippers and fish eggs**

BREEDING
**Age at first breeding: 1 year; breeding
season: probably April–July; number of
eggs: 1 or 2; incubation period: not known;
fledging period: not known; breeding
interval: 1 year**

LIFE SPAN
Not known

HABITAT
**Coastal mudflats, beaches, sand dunes,
coral reefs, reef flats and lagoons, mainly
in intertidal zone and never more than
⅔ mi. (1 km) inland**

DISTRIBUTION
**Coastlines of Indian Ocean, The Gulf and
Red Sea; breeds only in southern Red Sea,
Gulf of Aden and Gulf of Oman**

STATUS
Fairly common locally

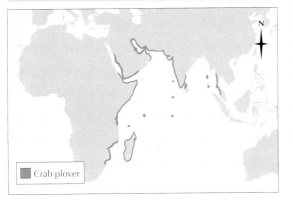

Crab plover

Nesting in holes

The breeding habits of crab plovers are unique among shorebirds, for they nest in burrows in sand dunes. It is during the breeding season that the crab plovers leave the shores and move a little way inland to banks and dunes protected from storms. Here the crab plovers make their burrows in small colonies, digging out the earth using their bill and feet. The burrows honeycomb the ground in the manner of rabbit warrens and may contain hundreds of nests.

Crab plover burrows resemble those made by auks and by some of the petrels. The opening, which may be on level ground or in a sandy cliff or bank, leads down at an angle, then curves upward and runs for about 5 feet (1.5 m) before ending in a nest cavity only a few inches beneath the surface. The floor of the tunnel is lower than the level of the nest cavity and thus may afford protection against flooding during rainstorms.

Within the nest cavity, on the bare ground, a single, large, goose-sized egg is laid; again, this is unusual for shorebirds, which usually lay two to four eggs. The incubation period has not, apparently, been recorded, neither is it known whether both parents incubate the egg. The chick hatches with a coat of down, as do the chicks of other shorebirds. It is able to run shortly after hatching but, unusually for a shorebird, stays in the nest cavity until fully-fledged, being fed by both parents on live crabs and other crustaceans.

Unusual egg for a shorebird

The egg of the crab plover is white, whereas those of other shorebirds and gulls are noted for their speckled black or brown colors. Those birds nest in the open and it is essential that their eggs are very well camouflaged and difficult to find. It is typical of birds nesting in tunnels or hollow trees to lay white eggs. In such situations camouflaged eggs are not necessary.

In proportion to the adult bird's size the crab plover's egg is very large, a quarter of the adult's weight of up to 14 ounces (400 g), in the case of females. To produce such an egg must place considerable stress on the female. As a general rule, proportionately large eggs are produced in species of bird in which the young leave the nest shortly after hatching; these species are known as nidifugous. A large egg with plenty of food for the chick inside allows the chick to develop to an advanced stage before hatching. Once hatched, the chick is fed by both parents. The crab plover is unusual in that its chick is hatched in an advanced state and yet it still remains within the nest cavity for some time. However, this behavior has a practical purpose: cavity-nesting birds are less vulnerable to predators than birds that use alternative types of nest.

CRABS AND LOBSTERS

CRABS ARE THE MOST ADVANCED members of the class Crustacea. Like all crustaceans, they have a hard external skeleton, numerous jointed limbs, segmented bodies and two pairs of antennae. Lobsters are a relatively ancient group of crustaceans compared to crabs; some fossils date from the early Jurassic Period, 200 million years ago. Along with most other commercially important crustaceans, such as crayfish, shrimps and prawns, crabs and lobsters belong to the order Decapoda, meaning "ten-footed."

The Decapoda is a highly diverse and successful order containing more than 8,500 species. In all decapods the head and thorax are fused and covered by a large shell called a carapace that extends down the sides of the body and encloses the gills. The carapace provides some protection from fish, octopuses, birds and other predators, and often also provides camouflage. The shells of the slipper lobsters of the genus *Scyllarides* are mottled and pinkish orange in color, enabling the lobsters to blend in with sand in tropical seas. The bodies of crabs consist mainly of a wide head and thorax, beneath which the abdomen is folded away.

The second major noteworthy feature of decapods is their large pincers or claws. These are highly mobile, powerful appendages used for feeding and sometimes during courtship. A decapod has five pairs of legs attached to its thorax, which it uses for walking. A pair of mandibles flanks or covers the mouth. These have grinding and biting surfaces. Behind the mandibles are two additional pairs of feeding appendages,

In common with all crabs and lobsters, the Sally Lightfoot crab, Grapsus grapsus, has a hard carapace, eyes on short stalks and five pairs of legs. The two foremost legs are powerful pincers.

CLASSIFICATION	
CLASS Crustacea	
ORDER Decapoda	
SUBORDER Reptantia	
INFRAORDER Brachyura Astacidea Palinura Anomura	
NUMBER OF SPECIES About 6,000	

Porcelain crabs (Petrolisthes) are able to move unharmed through the stinging tentacles of sea anemones, which protect the crabs from predators.

Tropical seas are home to by far the largest numbers of species and also to the greatest range of morphological variations.

True crabs

There are about 4,500 species of true crabs, in 47 families. True crabs belong to the infraorder Brachyura, meaning "short tails," a reference to their small abdomens. Brachyura are characterized by a flattened, heavily calcified carapace, ventrally fused in front of the mouth. The first pair of legs are developed as large claws, called chelae. The shell width of true crabs ranges from 7 millimeters in adult pigmy crabs of the genus *Sirpus* to over 16 inches (40 cm) in the Japanese giant spider crab, *Macrocheira kaempferi*) and 17 inches (43 cm) in the giant Tasmanian crab, *Pseudocarcinus gigas*. A Japanese giant spider crab has a claw spread of 11½ feet (3.5 m), but the Tasmanian species is heavier.

Lobsters and crayfish

True lobsters, including the common European lobster, *Homarus gammarus*, and the American lobster, *H. americanus*, belong to the infraorder Astacidea. Spiny lobsters and fresh-

the maxillae, which direct food into the mouth. Decapods also possess a balance organ, situated inside a cavity in the basal segment of each of the front antennae and linked to the outside by a slit in the shell.

Crabs, lobsters and crayfish are classed in the suborder Reptantia, or walking decapods, which is further divided into four infraorders. The species in these infraorders are mainly marine and estuarine. Some crabs can tolerate freshwater habitats but must return to the sea to breed. Truly freshwater representatives of the Reptantia include the freshwater crayfish and a few species of river crab. Terrestrial representatives are limited to several crayfish, which inhabit damp soil, and a small number of crabs, such as the robber, or coconut, crab, *Birgus latro*. Some crayfish are cave-dwellers.

The Reptantia are widely distributed from the intertidal, or littoral, zone along coastlines, through coastal waters and out to the deep sea. However, they are most abundant in the coastal shallows, where they are adapted to live in a variety of habitats.

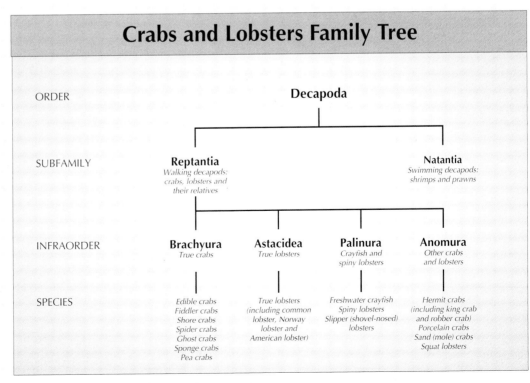

Crabs and Lobsters Family Tree

ORDER			Decapoda		
SUBFAMILY		**Reptantia** *Walking decapods: crabs, lobsters and their relatives*			**Natantia** *Swimming decapods: shrimps and prawns*
INFRAORDER		**Brachyura** *True crabs*	**Astacidea** *True lobsters*	**Palinura** *Crayfish and spiny lobsters*	**Anomura** *Other crabs and lobsters*
SPECIES		Edible crabs Fiddler crabs Shore crabs Spider crabs Ghost crabs Sponge crabs Pea crabs	True lobsters (including common lobster, Norway lobster and American lobster)	Freshwater crayfish Spiny lobsters Slipper (shovel-nosed) lobsters	Hermit crabs (including king crab and robber crab) Porcelain crabs Sand (mole) crabs Squat lobsters

water crayfish belong to the infraorder Palinura. All have an elongated abdomen, enclosed within a heavily calcified exoskeleton, and a well-developed tail fan. Unlike lobsters, the carapace of crayfish is fused ventrally with a solid plate. In lobsters, the first three pairs of limbs bear claws, the first set being larger than the following two, especially in males. The American lobster may weigh over 50 pounds (23 kg) and exceed 2 feet (60 cm) in length.

Other species

The infraorder Anomura is a very diverse group that includes the hermit crabs, porcelain crabs, sand crabs and squat lobsters. Squat lobsters are a small group of lobster-like crustaceans in which the abdomen is bent under the carapace. The most abundant members of this infraorder are hermit crabs, of which there are more than 600 species in seven families. Hermit crabs have soft bodies and protect themselves by taking over the abandoned shells of gastropods (marine snails), such as whelks. Most hermit crabs are marine, but there is a variety of terrestrial species, one of which lives inside hollow bamboo stalks and reeds.

The robber crab and the king crab have somewhat calcified abdomens as adults and do not inhabit gastropod shells. The porcelain crabs of the family Porcellanidae are superficially similar to true crabs but are in fact more closely related to squat lobsters.

Growth

The hard exoskeleton of crabs and lobsters represents up to 40 percent of their total body weight. This skeleton is non-living and, once hardened, it cannot change shape. It follows that the only way in which decapods are able to grow is by molting, a procedure also referred to as ecdysis.

A molting crab typically finds somewhere safe to hide and then deliberately cracks its shell by absorbing a large volume of water. It carefully withdraws its body from the old shell and continues taking in water to stretch the new one to the required size. The replacement shell takes between a few days and 2 weeks to fully harden. One species of shore crab, *Carcinus meanas*, is known to molt as many as 18 times during its 4-year life span. Many species of crab and lobster undergo a terminal molt, after which they can no longer grow. Others never stop molting completely, although the interval between molts may become extended. All decapods have the ability to regenerate a lost leg in the next molt.

A number of lobsters are very long-lived. Some species live for 10 years and the American lobster has a life span of 15–20 years; a few individuals survive for almost 100 years.

Reproduction

Female crabs, lobsters and crayfish generally mate soon after molting, when their skeleton is still soft. Certain spider crabs, in which copulation takes place when the shell has hardened,

Many of the true lobsters have large pincers, which are used for feeding and during courtship rituals. Those of the American lobster are strong enough to crush the shells of its main prey: crabs and clams.

Crabs are mostly omnivorous animals and take a wide range of food, including seaweed, worms, mollusks, clams and carrion. This red mangrove crab, Sesarma meinerti, *has emerged from its hole to feed on a mangrove leaf, its staple food.*

are an exception to this rule. The fertilized eggs are carried by the female on her underside until they are ready to hatch. The larvae swim freely and, before turning into juveniles, undergo several stages of development. These are (in order): prezoea, zoea and megalopa.

Locomotion

Crabs have eight walking legs distributed on a short body, and find it more convenient to move sideways. The ghost crab, *Ocypode cursor*, is particularly swift, reaching speeds of up to 5¼ feet (1.5 m) per second. The running movement places a great strain on the ghost crab's limbs. To alleviate this, the ghost crab alternates legs by abruptly swiveling through 180° without stopping, thus sharing out the strain evenly on all of of its legs. Most adult crabs do not swim well. An exception is the blue edible crab, *Callinectes sapidus*, of the Atlantic coast of North America, which can swim sideways, backward and forward. Lobsters usually move by crawling, though they can swim backward rapidly to escape predators.

Some decapods perform long migrations. Foremost among the migrant species are the spiny lobsters of the genus *Palinurus*, which may travel more than 60 miles (95 km) during a single migration. Gatherings of several hundred lobsters are not unusual at these times.

Feeding strategies

Crabs are primarily scavengers, feeding on plant and animal remains. Fiddler crabs of the genus *Uca* remove organic material from sand and mud with their mouthparts, depositing the indigestible particles as small balls. Many crabs are also active predators of fish, mollusks and other crustaceans. Some species, such as the tiny pea crabs of the family Pinnotheridae, obtain food by exploiting various marine invertebrates without killing them. Pea crabs are so named because they are

small enough to dwell inside the body cavities of other animals, like peas in a pod. Favorite hosts include bivalve mollusks, snails, sea cucumbers and sand dollars. The pea crabs steal food by straining it from their hosts' filtering apparatus. Such one-sided relationships between different animals are known as commensal relationships. Perhaps the most unusual diet of any decapod is that of the robber crab, a terrestrial species found near coasts in the Indian Ocean and South Pacific. It feeds exclusively on the nutritious pulp of broken coconuts and fallen fruits.

Crabs and lobsters play a vital ecological role in many marine ecosystems. They remove large volumes of decaying plant and animal matter from the shore and seabed, and convert this organic food energy into their own body tissues. The crabs and lobsters in turn provide food for many species of fish, otter, seal, seabird and octopus.

Conservation

Crabs, lobsters, crayfish and spiny lobsters are heavily exploited worldwide as human food. For example, the edible crab, *Cancer pagurus*, common lobster and Norway lobster, *Nephrops norvegicus*, are all popular dishes in northern Europe. One ancient Egyptian temple is decorated with a 3,500-year-old motif of the common spiny lobster, *Panulirus pencillatus*, suggesting that it has long been an important catch; the species is still a valuable shellfish today. Some species of crab and lobster, including certain spiny lobsters, are becoming scarce in some waters due to overfishing.

For particular species see:
- CRAYFISH • EDIBLE CRAB • FIDDLER CRAB
- HERMIT CRAB • KING CRAB • LOBSTER
- ROBBER CRAB • SHORE CRAB
- SPIDER CRAB • SQUAT LOBSTER

Index

Page numbers in *italics* refer to picture captions.
Index entries in **bold** refer to guidepost or biome and habitat articles.

Page numbers in *italics* refer to picture captions. Index entries in **bold** refer to guidepost or biome and habitat articles.